# 63 Advances in Polymer Science
Fortschritte der Hochpolymeren-Forschung

# Polydiacetylenes

Editor: Hans-Joachim Cantow

With Contributions by
H. Bässler, V. Enkelmann, H. Sixl

With 87 Figures and 11 Tables

Springer-Verlag Berlin Heidelberg GmbH 1984

ISBN 978-3-662-15239-3     ISBN 978-3-540-38951-4 (eBook)
DOI 10.1007/978-3-540-38951-4

Library of Congress Catalog Card Number 61-642

Typesetting: Th. Müntzer, GDR; Offsetprinting: Br. Hartmann, Berlin;

2154/3020–543210

# Foreword

Research on polydiacetylenes is not only an integral part of modern polymer chemistry but has attracted workers from many different areas of science as well thus creating a truly interdisciplinary field of many active developments. It has a number of quite different facets extending from preparative and mechanistic polymer chemistry via quantum chemistry and spectroscopy to materials science and biomimetic chemistry. The origin of all this activity can be traced to a single publication [1] by which the solid-state reactivity of certain substituted diacetylenes (*1*) was explained in terms of a polymerization reaction. The unique feature of this reaction is that it proceeds within the perfect lattice and, being completely controlled by the packing of the monomer, leads to perfect crystals of the corresponding polymer in a single crystal to single crystal type transition. Thus, for the first time, macroscopic and perfect single crystals of polymers could be prepared [1-3].

Although interesting within the framework of polymer physics and material science this would not be sufficient to attract so many workers from areas outside of conventional polymer research. Additional interest arouse because of the unusual structure of the polymers obtained via solid-state polymerization of diacetylenes and because of the mechanistic features related to its formation. Polydiacetylenes exhibit a fully conjugated and planar backbone in the crystalline state and are thus considered the prototype study object as far as the nature and physical behavior of polyconjugated macromolecules are concerned [4-6]. Theoretical discussions of the electronic structure of these polymers (*2*) lead to a description in terms of a wide band one-dimensional semiconductor [7].

$$R \, (e.g.) \!-\! (CH_2)_n \!-\! O \!-\! SO_2 \!-\!\!\!\!\langle \rangle\!\!\!\!-\! CH_3$$

$$-\!(CH_2)_n\!-\!CH_3 \qquad\qquad n \geqq 1$$

$$-\!(CH_2)_n\!-\!O\!-\!CO\!-\!NH\!-\!Ph \qquad\qquad -\!(CH\!\!=\!\!CH)_n\!-$$

$$3$$

Polyacetylene (3) which seems to be the simplest molecule exhibiting the feature of polyconjugation is, in reality, by far more complicated because it cannot be prepared in a single crystal-like texture; consequently the electronic phenomena of interest as for example optical spectra, conductivity and redox phenomena cannot be investigated without considerable ambiguity and are usually dominated by the defect structure of the sample rather than by the intrinsic nature of the molecule [8, 9].

Unlike polyacetylene, polydiacetylenes cannot be "doped", i.e. oxidized or reduced to give a polymeric salt of metal-like conductivity. Nevertheless, polydiacetylenes are interesting materials for possible applications in electronic devices.

They show very high carrier mobilities ($\mu_e \geqq 10^3$ cm$^2$ v$^{-1}$ sec$^{-1}$) in chain direction, when the carriers are injected from the electrode [10, 11].

The third order non-linear susceptibility coefficient $\chi(3)$ of polydiacetylenes has been found not unexpectedly very high and comparable to that of GaAs below the absorption edge [12]; moreover, switching times in non-linear absorption experiments of the order of $10^{-14}$ sec have been reported [13] giving polydiacetylenes a prominent place among the materials presently discussed in terms of effects like self-focusing, self-trapping, phase conjugation, optical bistability etc. which are fundamental to all-optical signal processing applications.

Another facet of polydiacetylene research grew out from the successful attempt to polymerize tenside molecules which had the diacetylene group as part of the hydrophobic section at the air-water interphase of a Langmuir-trough or in form of Langmuir-Blodgett-layers after transfer from the air-water interphase to a solid substrate [14, 15]. Layered polymer structures with molecularly controlled thickness are thus available. These structures are currently being investigated with respect to the above mentioned electronic properties in electronic and/or all-optical signal processing devices, in addition their use as resist materials in the fabrication of integrated electronic devices ("chips") has been discussed [16]. The observation that tenside-like diacetylene can be polymerized in surface or layered structures suggested that they could be polymerized in micellar or vesicular structures as well. This was shown to be the case by H. Ringsdorf and co-workers [17]. He and other workers demonstrated that micelles and vesicles build from diacetylenes can be permanently stabilized by photopolymerization. Incorporated into suitable phospholipid structures they can be used for reconstitution experiments and certain functions of the cell membrane and membrane-enzyme interaction can be studied using them as models. Finally, Chapman [18] showed that such diacetylenes can even be incorporated into the cell membranes of microorganisms with subsequent polymerization. These investigations have triggered a considerable activity of further polydiacetylene research in the area of biophysics and biochemistry.

The basic chemistry of diacetylene polymerization was developed by joint efforts of chemists and solid-state physicists. The very fact that the chain growth takes place inside the perfect lattice allows for the study of the intermediates of the reaction by combinations of ESR, NMR and optical spectroscopy with the molecular axes oriented with regard to the probing field direction via the crystal orientation. Thus, in combination with the results from X-ray structure analyses on the monomers and corresponding polymers the mechanistic details of the polymerization could be worked out. The active chain ends have initially the character of a radical but are better described as a carbene at degrees of polymerization $P_n \geq 6$. It can safely be stated, that the polymerization of diacetylenes is better investigated and understood than any other polymerization reaction at present, considering the insights into formation, electronic structure and reactivity of the intermediates.

It is the main purpose of the following articles by V. Enkelmann, H. Bässler and H. Sixl to review the present status of polydiacetylene research from the point of view of structure and reactivity including all the details known on the mechanism of polymerization of various diacetylenes. The material science aspects will not be treated to the same depth with exception of the photopolymerization and its possible application in the contribution of H. Bässler. Similarly, the solution properties of polydiacetylenes are not touched upon. The interested reader is refered to the current literature where the problems encountered when studying the solutions and the recrystallization behaviour of polyconjugated macromolecules have just started to be discussed adding a new chapter to the statistical mechanics and hydrodynamic behaviour of macromolecules [19, 20].

The field of polydiacetylenes, after 15 years of active research, is still very intriguing and full of surprises. It is now considered to be one of the classical fields by which organic solid state chemistry is developing both experimentally and conceptually [21]; it is foreseen to grow strongly in the coming years due to the interdisciplinary aspects of the field and its relation to the further development of modern technology and the molecular sciences.

Freiburg, March 1984                                   G. Wegner

# References

1. Wegner, G.: Z. Naturforsch. *24b*, 824 (1969)
2. Kaiser, J., Wegner, G., Fischer, E. W.: Israel J. Chem. *10*, 157 (1972)
3. Wegner, G.: Makromol. Chem. *145*, 885 (1971)
4. Bloor, D., Ando, D. J., Preston, F. H., Stevens, G. C.: Chem. Phys. Lett. *24*, 407 (1974)
5. Bloor, D.: "Experimental Studies of Polydiacetylenes: Model Conjugated Polymers" in "Quantum Theory of Polymers", Springer Lecture Notes in Physics *113*, 14 (1980)
6. Bloor, D.: in "Developments in Crystalline Polymers, Basset, C. D., ed. Appl. Sci. Publ., Englewood, N.J. Vol. 1 (1982), p. 151 ff.
7. Karpfen, A.: J. Phys. C *13*, 5673 (1980) and References therein
8. Ito, T., Shirakawa, H., Ikeda, S.: J. Polymer Sci., Polym. Chem. Ed. *12*, 11 (1974)
9. Wegner, G.: Angew. Chem. Int. Ed. Engl. *20*, 361 (1981)
10. Donovan, K. J., Wilson, E. G.: Phil. Mag. *B44*, 9 (1981)
11. Spannring, W., Bässler, H.: Chem. Phys. Lett. *84*, 54 (1981)
12. Sauteret, C., Hermann, J. P., Frey, R., Pradiere, F., Ducuing, J., Baughman, R. H., Chance, R. R.: Phys. Rev. Lett. *36*, 956 (1976)
13. Smith, P. W.: The Bell System Technical J. *61*, 1975 (1982)
14. Tieke, B., Wegner, G., Naegele, D., Ringsdorf, H.: Angew. Chem. Int. Ed. Engl. *12*, 764 (1976)
15. Tieke, B., Graf, H. J., Wegner, G., Naegele, B., Ringsdorf, H., Banerjie, A., Day, D., Lando, J. B.: Colloid and Polymer Sci. *255*, 521 (1977)
16. Compare all References in Thin Solid Films *99* (1983) (Special Issue on LB-Layer Structures)
17. Gros, L., Ringsdorf, H., Schupp, H.: Angew. Chem. *93*, 311 (1981)
18. Johnston, D. S., Sanghera, S., Pons, M., Chapman, D.: Biochim. Biophys. Act. *602*, 57 (1980)
19. Lim, K. C., Fincher, L. A., Heeger, A. J.: Phys. Rev. Lett. *50*, 1934 (1983)
20. Müller, M. A., Schmidt, M., Wegner, G.: Makromol. Chem. Rapid Commun. *5*, 83 (1984)
21. Compare "Proc. of the VI. Int. Conf. Chem. Org. Solid State", Freiburg, 1982, Mol. Cryst. Liq. Cryst. Vol. *93* and *96* (1983)

# Editors

# Editorial

With the publication of Vol. 51, the editors and the publisher would like to take this opportunity to thank authors and readers for their collaboration and their efforts to meet the scientific requirements of this series. We appreciate our authors concern for the progress of Polymer Science and we also welcome the advice and critical comments of our readers.

With the publication of Vol. 51 we should also like to refer to editorial policy: *this series publishes invited, critical review articles of new developments in all areas of Polymer Science in English (authors may naturally also include works of their own).* The responsible editor, that means the editor who has invited the article, discusses the scope of the review with the author on the basis of a tentative outline which the author is asked to provide. Author and editor are responsible for the scientific quality of the contribution; the editor's name appears at the end of it.
Manuscripts must be submitted, in content, language and form satisfactory, to Springer-Verlag. Figures and formulas should be reproducible. To meet readers' wishes, the publisher adds to each volume a "volume index" which approximately characterizes the content.

Editors and publisher make all efforts to publish the manuscripts as rapidly as possible, i.e., at the maximum, six months after the submission of an accepted paper. This means that contributions from diverse areas of Polymer Science must occasionally be united in one volume. In such cases a "volume index" cannot meet all expectations, but will nevertheless provide more information than a mere volume number.

From Vol. 51 on, each volume contains a subject index.

Editors                                                         Publisher

# Table of Contents

# Photopolymerization of Diacetylenes

H. Bässler
Fachbereich Physikalische Chemie, Philipps-Universität, Hans-Meerwein-Straße, D-3550 Marburg, FRG

*This article reviews the current state of art of photopolymerization of crystalline diacetylenes, although complementary information on thermal and γ-polymerization is also included whenever felt appropriate. Priority is given to an outline of recent achievements regarding reaction kinetics and energetics as well as to model considerations for both initiation and propagation of the polymer chain. It will be shown that at least for diacetylene-bis(p-toluenesulfonate) (TS) existing experimental data allow establishing a consistent picture of the polymerization process. After discussing recent works on photopolymerization in Langmuir-Blodgett-multilayer assemblies, including sensitization and self-sensitization effects, potential applications of the unusual optical, electrical, and reaction properties of diacetylenes are outlined.*

# 1 Introduction

Although it has been known since more than a century that certain crystalline diacety-lenes undergo a dramatic color change upon prolonged storage under ambient conditions, it was not until Wegner's [1-5] interpretation of the effect that these materials became popular objects of chemical and physical research. Based upon the principles of topochemical reactions developed by G. M. L. Schmidt and co-workers [7] he concluded that the color change must be the result of a polymerization process in course of which the C1 and C4 carbon atoms of adjacent diacetylene moieties in a molecular stack are linked together. According to the following scheme elongated polymer chains are formed under preservation of the single crystalline phase structure provided that the molecular motions accompanying the chemical transformation compensate each other in a way as to minimize the overall changes of the crystallographic parameters [8-11]. It is also obvious that the reaction must be encumbered by an energy barrier. No matter what the total energy of the polymeric reaction product relative to the monomer is, mutual approach of the reaction centers must involve a molecular rotation which has to act against the repulsive forces exerted by the adjacent C1 and C4 atoms before bond formation. This explains why the reaction usually requires excess energy supplied by UV-photons, X-ray or γ-quanta to get started. In some cases phonons generated thermally are sufficient to trigger the polymerization process.

monomer single crystal            polymer single crystal

Not only does the solid state polymerization of crystalline diacetylenes offer a unique tool for studying topochemical reactions, it also affords fascinating, if special-ized model systems for polymerization reactions in general. Of particular advantage is the absence of disorder [12], always present in homogeneous solution reactions and often obscuring analysis of the individual reaction steps. One can therefore apply high resolution optical [13-17] and magnetic [18-25] resonance techniques to elucidate the reaction mechanism. Such studies have yielded an unprecedented wealth of infor-mation about the individual reaction steps and the reaction intermediates which may be of relevance for other systems as well.

The reaction scheme shown above also allows prediction of some of the prominent properties of polydiacetylenes. Increasing the conjugation length of a molecule lowers the energy of elementary excitations of a conjugated chain [26, 27]. Therefore, the highest filled molecular π-orbitals, i.e., the valence band of the chain, must experience an upward and the lowest empty π-orbitals, i.e., the conduction band, a downward shift in energy upon polymerization and the energy of the optical transition must decrease. As a consequence, one can expect that it is the electronic structure of the chain rather than that of the substituents which controls the electronic properties of the polymer, quite in contrast to the situation encountered with conventional polymers having σ-bonded backbones where the substituents play the dominant role. In particular, the large degree of anisotropy brought about by the different types of bonding parallel and perpendicular to the chain should be revealed in electronic transport properties. In fact, crystalline diacetylenes have been shown to provide the closest approach to one-dimensional systems known so far and are therefore widely used as model systems to study effects of reduced dimensionality on both electrical [28-34] and optical [35-40] properties. For a more detailed survey of the properties of polydiacetylenes the reader is referred to several review articles [41-47].

The aim of this article is to provide an overview of the polymerization of diacetylenes. The focus will be on optical excitation, although some results on thermal reactivity will also be quoted to illustrate analogies. Comprehensiveness is not intended, instead, emphasis will be placed on model considerations. Structural aspects of the polymerization process, as well as the low temperature spectroscopy of reaction intermediates will only briefly be addressed since they are treated in detail in the contributions of V. Enkelmann and H. Sixl in this volume.

## 2 Phenomenological Aspects of the Polymerization Process

Upon exposure to heat, UV- or γ-radiation diacetylenes are converted from a soluble monomer crystal which is transparent if pure, i.e., free of residual polymer, to a deeply colored polymer crystal. With a few exceptions, the latter is insoluble in all common solvents. The color arises from the lowest π-electron transition of the conjugated polymer backbone, which has its maximum near 600 nm. It is of excitonic origin [29, 48-50] and carries an oscillator strength of the order unity. Both insolubility and optical absorption can be used to monitor the degree of conversion as a function of reaction time. In the first case the crystal is dispersed mechanically and the amount of polymer is determined gravimetrically from the weight fraction of insoluble material. Because of the high absorption coefficient of the polymer crystal, which is in the order $10^6$ cm$^{-1}$ for light propagation perpendicular to the polymer chain and polarized parallel to the chain, and $\sim 2 \cdot 10^4$ cm$^{-1}$ for perpendicular polarization [36, 37], direct absorption spectroscopy is limited to ultra-thin samples or to very low conversions. However, Chance and Sowa [51] have demonstrated that at least for TS (for compound symbols see Table 1) reliable conversion studies do not require single crystals but can be performed with microcrystalline samples. This allows application of the diffusive reflectance technique for optical detection of the polymer content.

**Table 1.** List of symbols and structure formula of the substituents for the diacetylenes treated in the text

| Symbol | Substitutent (s) |
|---|---|
| TS-6 (or TS) | $-CH_2-O-SO_2-\langle\bigcirc\rangle-CH_3$ |
| TS-12 | $-(CH_2)_4-O-SO_2-\langle\bigcirc\rangle-CH_3$ |
| MBS | $-(CH_2)-O-SO_2-\langle\bigcirc\rangle-OCH_3$ |
| TCDU | $-(CH_2)_4-O-CO-NH-\langle\bigcirc\rangle$ |
| EUHD | $-(CH_2)-O-CO-NH-C_2H_5$ |
| HDU | $-(CH_2)-O-CO-NH-\langle\bigcirc\rangle$ |
| MCD | $-(CH_2)_4-O-CO-NH-CH_3$ |
| n-BCMU | $-(CH_2)_n-O-CO-NH-CH_2-CO-O-(C_4H_9)$ |
| DCH | $-CH_2-N\langle\text{carbazole}\rangle$ |
| TCDA | $R_1: -(CH_2)_9-CH_3$ <br> $R_2: -(CH_2)_8-COOH$ |

It is generally agreed that polymerization of diacetylenes is a multistep process. Initially a reactive center is created which subsequently grows by addition of monomer molecules. Formally the reaction can be described by a first order rate equation [52]:

$$\frac{dX}{dt} = K(1 - X) \tag{1}$$

where $1 - X$ and $X$ are the fractional monomer and polymer content, respectively. The rate constant K is the probability $\gamma$ that a reaction center is created per unit time multiplied by the number of monomers, n, that react per initiation event. In case of thermal polymerization:

$$K = n\gamma_0 \exp\left(-\frac{E_a^{therm}}{kT}\right), \tag{2}$$

$E_a^{therm}$ being the thermal activation energy. In case of optical generation of the reaction center, $\gamma$ is the probability that a monomer molecule absorbs a photon, multiplied by the probability q that excitation will lead to chain initiation. If the thickness (d) of the absorbing crystal is small compared to the penetration depth ($\alpha^{-1}$) of the exciting radiation:

$$K = (I_0\alpha/N_0)\, nq \tag{3}$$

where $I_0$ is the incident photon flux (photons/cm² s) and $N_0$ is the number of molecules per unit volume. If $d > \alpha^{-1}$, the factor exp ($-\alpha x$) has to be incorporated in Eq. (3) where x is the direction perpendicular to the irradiated surface located at x = 0. Then both K and X depend on the position of the absorbing molecule inside the absorber and the reaction becomes spatially inhomogeneous.

Eqs. (1) to (3) indicate that conversion studies under conditions where thermal polymerization prevails can only yield $E_a^{therm}$ and the product $n\gamma_0$, whereas the photon-induced reaction provides information on the product nq. To disentangle chain initiation and chain propagation effects an independent determination of the kinetic chain length is required.

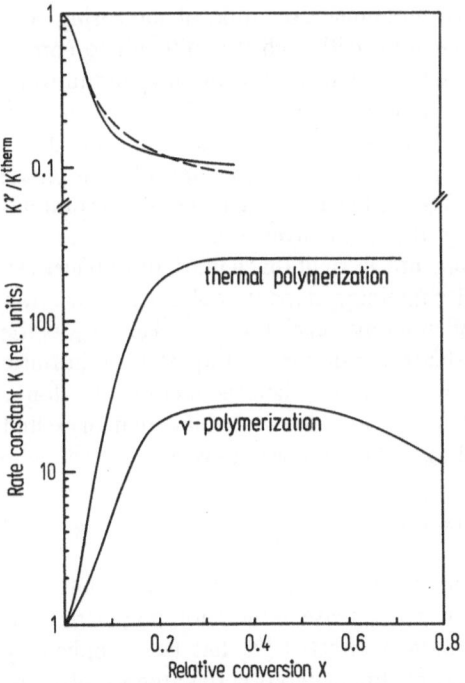

**Fig. 1.** Lower part: Rate constant for thermal and $\gamma$-polymerization of TS-6 as a function of conversion normalized to the rate constant at X → 0. Curves are calculated from published time-conversion curves according to $K = (1 - X)^{-1}\, dX/dt$ where X is the relative polymer content. $\gamma$-Polymerization data are from Ref. [53], thermal dates represent an average of literature dates, see e.g., [52]. Upper portion: $K^\gamma/K^{therm}$ vs. conversion. The dashed curve is calculated on the basis of the energy transfer model of Ref. [66] for a quenching constant $k_q = k_t\tau_0 = 35$

Representative for systems exhibiting sigmoidal conversion curves Fig. 1 shows experimental results for the rate constant of the reaction of TS, evaluated from thermal and γ-polymerization data [53] according to $K = (1 - X)^{-1} dX(t)/dt$, and normalized to the rate constant in the low conversion limit. It is obvious, that at low conversion K depends on X, contrary to what is to be expected for a simple first order reaction. The functional form of K(X) is different for the two modes of polymerization. The overall increase of K with increasing X reveals an autocatalytic reaction enhancement. A measure for its efficiency is the ratio $K(X = 0.5)/K(X = 0)$ which turns out to be about 200 for TS under thermal polymerization conditions. This effect is often observed with disubstituted diacetylenes [54−58], albeit with different kinetic parameters, and indicates that the conditions for the reaction to proceed are improved as more polymer is formed. This point will be further elaborated in the following chapters. There are, however, systems that do not exhibit autocatalytic reaction enhancement. Among these are TCDU [59], MBS [60, 61], urethane substituted diacetylenes, notably 3-BCMU and 4-BCMU [62], and multilayer systems [63].

There is an early report in the literature [2] claiming absence of the autocatalytic reaction enhancement in TS if the reaction is induced by UV-excitation of the monomer crystal. The implication would be that thermal and UV-polymerization involve different mechanisms. Later on, however, Chance and Patel [53] found this to be an artifact caused by the neglect of spatially inhomogeneous absorption by polymer molecules which effectively competes with monomer excitation at increasing conversion and prematurely terminates the reaction. Although it is difficult to correct X(t)-curves obtained under UV-excitation for polymer absorption quantitatively, particularly if irradiation is done with unpolarized non-monochromatic light, it turns out that there is a qualitative agreement between X(t)-curves obtained under γ- and UV-irradiation. Application of this correction, however, does not solve the puzzle why in case of γ- or UV-polymerization of TS, the reaction rate increases less dramatically with conversion, than observed upon thermal conversion.

The latter effect can be explained by taking into account that a partially polymerized diacetylene crystal is a molecular crystal containing dopant molecules with lower lying optical transition acting as traps for monomer excitations [64]. Take $k_{tr}$ as the rate constant for non-radiative energy transfer from a donor to a trap, $\tau_0$ as the intrinsic donor lifetime, and $c_t$ as the relative trap concentration, then the lifetime of a donor in presence of traps would be $\tau = (\tau_0^{-1} + k_{tr}c_t)^{-1}$. Since the probability that an excited donor can initiate a chain is proportional to its lifetime, we arrive at:

$$q(X)/q(X = 0) = [1 + (k_{tr}\tau_0) X]^{-1} \qquad (4)$$

for the relative chain initiation probability in presence of a relative polymer content X. This concept, originally proposed to explain the decrease in the photopolymerization yield in a system where spectral sensitization is effective [65], has been applied by Prock et al. [66] to 4-BCMU. The advantage of this material is the absence of auto-catalytic reaction enhancement. Therefore, the only concentration effect acting on the rate constant K in course of a UV-photopolymerization experiment should be quenching of the excited monomer state by the polymer provided that the sample thickness is small enough to render effects of inhomogeneous sample absorption negligible. In fact, Fig. 2 demonstrates that excited state quenching accounts very well for the decrease of the conversion rate with increasing conversions. The fit parameter, $k_{tr}\tau_0$, turns out to be 20. Inserting $\tau_0^{-1} = 10^8 \ldots 10^9 \text{ s}^{-1}$ 1 gives $k_{tr} \sim \sim 10^9 \ldots 10^{10} \text{ s}^{-1}$. This is a surprisingly low value for a molecular crystal where energy transfer normally occurs via rapid exciton hopping [64], and suggests that single step Förster transfer is the dominant transfer mechanism. This conclusion concurs with Niederwald et al.'s [67] observation that in a chemically pure TS crystal no energy transfer to physical defects is detectable. Apparently the $S_1$ state of the diacetylene moiety is rapidly localized involving structural relaxation.

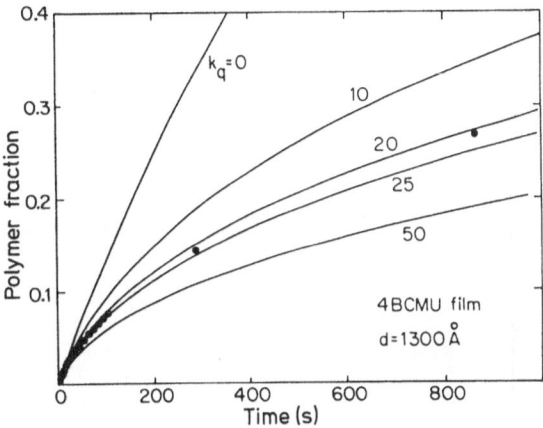

**Fig. 2.** Polymer fraction vs. irradiation time for a 4-BCMU film, 1300 Å thick. Data points are experimental, full curves are obtained for the energy transfer model for various quenching parameters $k_q = k_{tr}\tau_0$ (from Ref. [66])

The energy transfer concept can also be applied to explain why $K^\gamma(X)$ rises more slowly with conversion than $K^{th}(X)$ does. In Fig. 1 the ratio $K^\gamma(X)/K^{th}(X)$ is plotted versus X and compared with the prediction of the energy transfer concept assuming a quenching parameter $k_{tr}\tau_0 = 35$. The agreement is very satisfactory. The implication would be that thermal, UV-, and $\gamma$-polymerization only differ with respect to the chain initiation mechanism, the chain propagation mechanism being essentially the same. However, this conclusion may not be of general validity. Based on the strain dependence of the frequencies of backbone vibrations, Galiotis et al. [69] concluded that in a thermally polymerized ethylurethane diacetylene (EUHD), the chains are considerably longer than in X-ray polymerized samples. Conceivably, high energy radiation produces defects which interfere chain propagation. The importance of such an effect should decrease as additional chain limiting factors become dominant, e.g., lattice mismatch between monomer and polymer.

Unless proper account is taken for excited state quenching meaningful data regarding the quantum yield of photopolymerization can only be extracted from measurements under low conversion conditions, say $X \lesssim 0.03$. Owing to difficulties in

---

1 Recently Niederwald et al. [67] observed strong fluorescence emitted from a TS crystal, the time decay following the profile of a 10 ns excitation pulse. Combining their statement of high fluorescence yield with the fact that the singlet transitions of diacetylenes have low oscillator strength [68] suggests that $10^{-9} \lesssim \tau_0 \lesssim 10^{-8}$ s.

measuring absolute photon doses, quantitative experimental information is contradictory. Whereas Chance and Patel [53] derived nq = 1.5 from continuous irradiation experiments with an uncertainty of a factor of 5, Bhattacharjee and Patel [70] reported nq = 0.06 ± 0.03. The yield was found to be thermally activated with an activation energy $E_a^{opt}$ = 0.13 eV. From the change in optical density following excitation by a 308 nm pulse of an excimer laser Niederwald et al. [71] derived nq = 0.07 ± 0.02 at room temperature and a *negative* temperature coefficient. Recent work in this laboratory [72] gave nq = 0.15 ± 0.05 at $\lambda_{ex}$ = 308 nm with a tendency to increase upon lowering $\lambda_{ex}$ (see Fig. 3). In summary, it seems fair to conclude that nq ~ 0.1 represents a reasonable order of magnitude estimate for TS-6 in the low conversion limit.

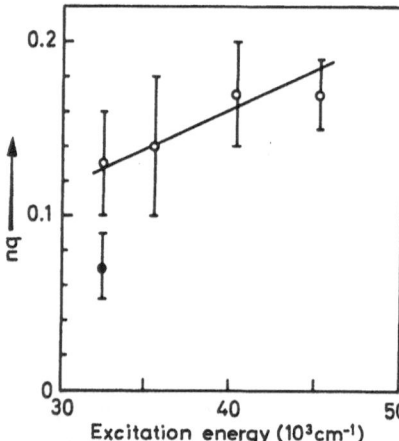

Fig. 3. Total polymer yield (nq) for TS-6 as a function of incident photon energy. (Open circles are from Ref. [72], the full circle is from Ref. [71])

For 4-BCMU, both photoacustic [73] and thin film absorption [66] studies consistently show that nq must be of order 100 at 295 K. The most accurate value appears to be nq = 60, associated with an activation energy of 0.3 eV, which has been determined calorimetrically [74]. For mixed crystals of HD and Phenazine (HD-Ph) nq = 0.5 and $E_a^{opt}$ = 0.17 eV has been reported under excitation of the diacetylene moiety [65].

It is generally agreed that chain initiation is a statistical process occurring at random within the bulk of the monomer matrix and leading to a solid solution of polymer within the parent crystal [3, 51, 75]. Recent neutron scattering experiments on TS by Grimm et al. [76] did not show any indication of inhomogeneous polymer growth which might have delineated existence of specific nucleation sites. Chain interaction was only noted near 50 percent conversion. Polymerization may, however, give rise to a phase transition at high conversions. An example is DCH where Enkelmann et al. [55] observed a reaction-induced phase transition associated with a shearing of the monomer lattice. It improves molecular packing and consequently gives rise to autocatalytic reaction acceleration. A phase transition that nucleates at defects and destroys the macrocrystalline order of the matrix thereby rendering the reaction inhomogeneous was detected in TCDA above 50 percent conversion [59].

# 3 The Chain Length Problem

## 3.1 Experimental Results

Determining the length of the polymer product is of key importance for understanding the polymerization process and received considerable attention ever since the reaction was discovered. On the experimental side direct determination is impeded by the insolubility of most polydiacetylenes — exceptions will be mentioned below — which precludes application of standard methods for measuring the molecular weight.

Of particular interest are changes in the chain length occurring in connection with the autocatalytic reaction acceleration in TS-6. Numerous thermal polymerization studies showed that the activation energy is $E_a^{th} = 1.00 \pm 0.02$ eV, independent of conversion. Consequently the autocatalytic reaction enhancement cannot be the result of an increase of the Boltzmann factor. Instead, an increase in the number of monomers consumed per primary chain initiation event has been postulated. Experimentally $n(X = 0.5)/n(X = 0) \simeq 200$ is found [51].

An early estimate, based upon the position of the optical absorption of TS-polymer at low conversion gave $n(X = 0) \sim 24$ repeat units [51]. Although the conclusiveness of the procedure to derive this number is questionable — for $n \gtrsim 10$ the position of the lowest absorption band of the polymer backbone is almost independent of $n$ [17] yet sensitive to chain expansion, known to be important at low X [77, 78] — this number has turned out to be essentially correct. Analysing diffusive features in the X-ray diffraction pattern of partially polymerized TS crystals Albouy et al. [79] recently concluded that at room temperature $n$ increases smoothly from $n = 18$ at $X \simeq 0$ to $n = 30$ at $X \simeq 0.1$ (see Fig. 4). Brillouin scattering experiments by Enkelmann et al. [80] gave $n \geq 5$ for $X \leq 0.1$.

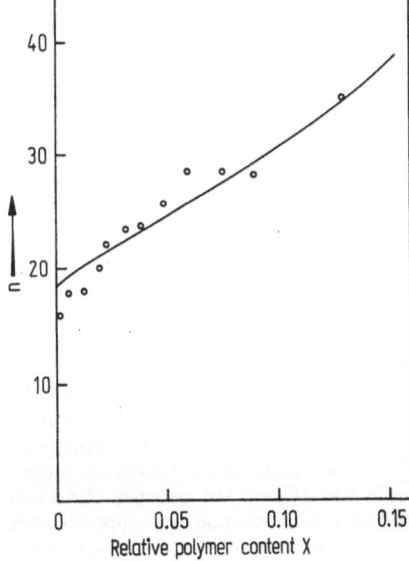

**Fig. 4.** Average number of repeat units contained in a TS-6 polymer chain as a function of conversion. Data points are experimental (Ref. [79]), the full curve is calculated from Eq. (9) in Section 6.1

Both ESR experiments [81] as well as viscosity measurements [2] are consistent with the existence of long polymer chains in high conversion TS. In an attempt to find a way for determining the chain length that does not require sample solubility, Mondong and Bässler [82] applied the technique of photopolymerization under spatially intermittent irradiation. This work was stimulated by Avakain and Merrifield's classic experiment on triplet exciton diffusion in crystalline anthracene. In this experiment the decrease of the triplet concentration caused by exciton diffusion into unirradiated crystal areas was inferred from changes in the intensity of the delayed fluorescence [83]. Adapting their set-up to the study of photopolymerization of TS a micro-mesh with square holes of typically 50 μm edge length and a bar width ranging from 7 to 30 μm was put on top of the (100)-face of a TS crystal which had been subjected to thermal polymerization up to a certain degree of conversion $(X \leq 0.15)$. The crystal was irradiated with a continuous UV light source and subsequently etched to remove unreacted monomer from the surface. The etch profile was recorded with a scanning electron microscope. Figure 5 presents an etch pattern obtained with a crystal originally containing $X = 0.15$ polymer, sufficient to establish autocatalytic reaction conditions for the subsequent photoreaction. The asymmetry

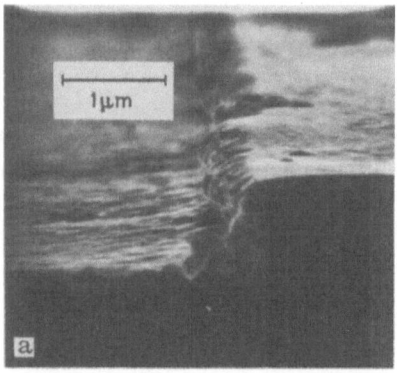

**Fig. 5a and b.** Etch profile of TS-6 crystals after thermal polymerization for 4 h (**a**) and 20 h (**b**), respectively at 60 °C, and subsequent photopolymerization for 2 h (**a**) and 30 min (**b**), respectively, at 20 °C. The width of the grid bars of the irradiation mesh was 20 μm. The polymer content at the beginning of irradiation was approximately 2 percent in case (**a**) and 15 percent in case (**b**)

of the etch pattern is in accord with the accepted view that UV-initiated polymer chains grow along [0 1 0]. Therefore, their heads can only penetrate into the unirradiated zones running along [0 0 1] and cause smearing out of the etch profile. Sharp profiles were only observed if the initial polymer content was well below the autocatalytic threshold concentrations. This experiment demonstrated that the autocatalytic reaction enhancement does involve an increase of the number of monomers converted per chain initiation event and suggested an ultimate average chain length of order 4 µm, equivalent to 8000 repeat units.

There is, however, serious doubt that 4 µm is the kinetic length of an individual chain. Consider a monomer crystal that contains a fraction $X_0$ of short chains consisting of $n_0$ repeat units. The average number of monomers enclosed between the ends of two polymer chains in a row and available for subsequent photopolymerization is:

$$\langle n \rangle = n_0(X_0^{-1} - 1) \tag{5}$$

Inserting $n_0 = 30$, $X_0 = 0.1$ gives $\langle n \rangle = 270$. $\langle n \rangle$ should be an upper limit for the average chain length in the autocatalytic reaction regime, if for reasons to be discussed below any further limitation of the kinetic chain length is suspended and if dead chain ends are not revived by recombination with a running chain.

Progress in this field has been aided by the discovery of soluble diacetylenes allowing a chromatographic study of the molecular weight distribution [84-86]. The results of the most recent work by Wenz and Wegner [87] on TS-12 are in accord with the above reasoning. TS-12 is a soluble analog to TS-6, differing only with respect to the number of $CH_2$-units in the substituents and likely to mimic the behavior of TS-6 reasonably well. Figure 6 shows that at low conversions, $X < 0.05$, only short chains with

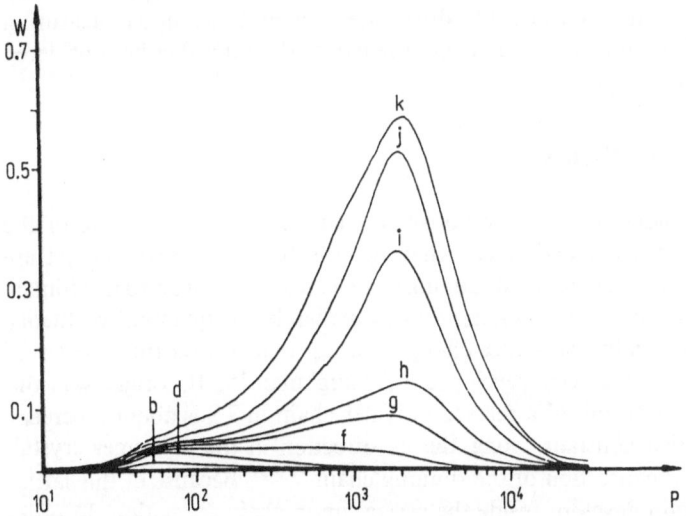

**Fig. 6.** Distribution of the average number of repeat units (P) contained in the polymer of TS-12 at various conversions. (b: X = 0.01, d: 0.025, f: 0.06, g: 0.12, h: 0.17, i: 0.35, j: 0.50, k: 0.64) (from Ref. [86])

$\langle n_0 \rangle = 60$ are formed. $X \simeq 0.05$ is the critical concentration at which in TS-12 autocatalytic reaction enhancement becomes important. With increasing X a second component appears in the molecular weight distribution displaying a peak near $\langle n \rangle = 2000$, virtually independent of conversion. It is readily verified that the data set $n_0 = 60$, $\langle n \rangle = 2000$, and $X_0 = 0.05$ is compatible with Eq. (5). Several important conclusions follow from this work: (i) The short chains formed during the initial stage of the reaction remain intact up to conversions of 70 percent indicating that they cannot be incorporated into longer chains; (ii) a simple statistical model seems to reproduce the basic features of the distribution of chain lengths [88]; (iii) the average maximum chain length is determined by the number of short chains produced during the induction period, and (iv) the autocatalytic behavior arises from the increase in the chain length rather than an increase of the number of chain initiation events as evidenced by the linear relationship between the number of chains formed and the irradiation dose.

There is no direct evidence that in TS-6 individual chains formed in the auto-catalytic reaction regime are only a few hundred repeat units long, contrary to what is to be expected from the 200-fold increase of the reaction rate. However, there is a recent estimate based on polarization currents measured during thermal polymerization. Following earlier work on the pyroelectric response of TS [89, 90], Bertault et al. [91] attributed the electric signal to an alignment of non-compensated dipolar defects associated with the chain ends. From the average value of the polarization they concluded that one chain can at most comprise a couple of hundred monomer units.

Knowledge of the chain length n and the overall yield nq allows evaluating the primary quantum yield q for optical chain initiation. Unfortunately, data are available for few systems only. Based on $\lim_{X \to 0} n \simeq 20$ [79] and $nq \simeq 10^{-1}$ (see above), $q \approx 5 \cdot 10^{-3}$ follows for TS-6. For 4-BCMU, a soluble diacetylene not exhibiting an autocatalytic reaction regime, $n = 2400$ [84] and $nq = 60$ (at 295 K) [74] have been reported, giving $q = 0.025$. In view of the considerable difficulties encountered upon measuring absolute doses in UV required to calculate nq, these data should rather be considered as order of magnitude estimates.

## 3.2 The Elastic Strain Theory

Simple statistical considerations of the above sort can provide an estimate of the maximum increase of the average kinetic chain length during the reaction, yet are unable to predict whether an individual running chain can actually grow to its ultimate length determined by the number of monomers available. A quantitative theory assessing the effects of matrix structure changes during reaction on the kinetics of the subsequent reaction has been developed by Baughman [52]. Its origin was the recognition that the repeat unit of a relaxed polymer chain in TS is about 5 percent shorter than the lattice constant along the $\vec{b}$ direction in the monomer crystal, which coincides with the direction of the running chain [75, 92]. Because of this lattice mismatch internal strain develops inside the crystal upon chain formation. In order that monomer and polymer be in register, chains have to be expanded and/or the monomer matrix has to be compressed. Which of these effects predominates depends on conversion and on the elastic moduli of monomer and polymer. At low X the matrix influence dominates forcing polymer chains to expand as evidenced by the blue shift of their optical absorption band [78]. Owing to their greater elastic stiftness the compressing effect of the polymer chains on the matrix takes over at comparatively low conversions to ultimately force the monomer lattice into register with the polymer product thereby improving the reaction conditions. At arbitrary conversion the lattice constant is a function of X, determined by the condition that the total free energy of the system be a minimum. By considering only the elastic energy term and ignoring structural changes perpendicular to the chain direction, Baughman derived an expression for the lattice parameter b(X) which meanwhile has been verified by both direct dilatometric recording during thermal polymerization of TS-6 [93] as well as X-ray analysis [79,80] (Fig. 7).

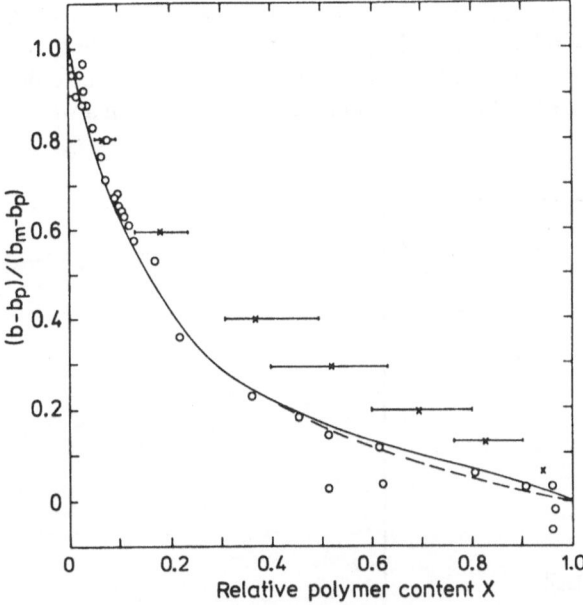

**Fig. 7.** Relative change of the lattice parameter in TS-6 along the chain direction as a function of conversion X. Indices m and p refer to monomer and polymer, respectively. The full curve is derived from a continuous dilatometric measurement [93], the dashed curve is calculated according to Ref. [52] by minimizing the elastic energy assuming a ratio $E_p/E_m = 5.0$ [95] for the elastic constants of polymer and monomer, respectively. Open circles are X-ray data from Ref. [79], crosses represent data from optical absorption studies [78]

The correlation between the effective strain of a polymer chain and its absorption energy, established from direct absorption measurements on stretched polymer fibres [77], allows an independent determination of the length of the repeat unit as a function of X via optical absorption measurements [78]. Data obtained by this method display a systematic deviation from the theoretical plot (Fig. 7) as do data derived from resonant Raman measurements [94]. This indicates that the agreement between experimental data and predictions of the elastic strain (ES)-theory is fulfilled only as long as the experimental conditions reflect the premises of the theory, notably its reduction to a strictly one-dimensional problem. Since electronic states of the polymer backbone as well as its vibrational modes reacts sensitively on both length of the overall repeat unit and readjustment of bond angles and bond lengths as a result of lateral forces acting on the chain, this condition is violated when performing optical chain detection experiments. Dimensional changes occurring along [1 0 0] and [0 0 1] in course of monomer-polymer conversion [92,93] are therefore reflected in both absorption and resonant Raman studies in a manner not predictable on the basis of the ES-theory.

The main goal for developing the ES-theory was to afford a framework for understanding the increase of the chain length within the autocatalytic reaction regime. Assuming that (i) both creation and deactivation of the active species initiating a chain is independent of conversion X, (ii) the probability that an active species initiates a chain is $\ll 1$ and (iii) the lifetime of reactive ends on growing chains is independent of X, Baughman calculated the relative chain length as a function of X. He found that in the conversion range $0.05 < X < 0.4$ n increases by approximately one order of magnitude. The result of the calculation is shown in Fig. 8 normalized in a way as

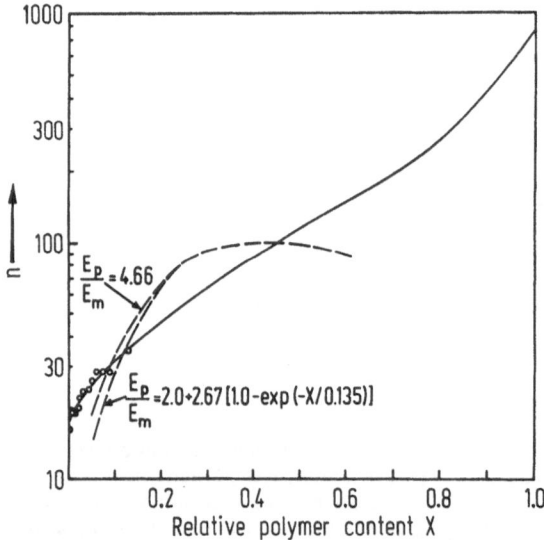

**Fig. 8.** Number of repeat units per polymer chain as a function of conversion. Dashed curves are from Ref. [52] and were calculated on the basis of the elastic chain theory for different ratios of the elastic constants $E_m$ and $E_p$ of monomer and polymer, respectively, and normalized to fit the experimental data near $X = 0.1$. The full curve is calculated according to Eq. (9) (Sect. 6.1) ignoring chain blocking effects. Data points are experimental [79].

to match Albouy et al.'s X-ray data for $X \simeq 0.1$. Extrapolating experimental low conversion data for n employing the ES-theory would predict a chain length of not more than 100 repeat units, equivalent to 500 Å, in the high conversion regime. The autocatalytic enhancement factor would then come out to be a factor of 3 less than the maximum enhancement factor resulting from statistical considerations. This result, if correct, would imply that in TS-6 the average length of a running chain is not determined by the number of monomer molecules available for reacting between the ends of existing chains, contrary to what has been found with TS-12.

Despite the success of the ES-theory to explain the dimensional changes occurring in TS-6 along [0 1 0] as a function of conversion it thus appears, that difficulties are encountered when it is applied to predict the increase of the chain length within the autocatalytic reaction regime in a quantitative fashion [95]. A similar reasoning was put forward by Lochner et al. [96] who analyzed the effect of hydrostatic pressure on the thermal reactivity of TS-6. They concluded that the decrease of the induction time with pressure occurred too fast to be explained in terms of the ES-theory alone. Further, the ES-theory implies that both chain propagation and initiation are similarly affected by a decrease of the elastic energy contribution to the total free energy change during the reaction. One would therefore expect the activation energy for thermal polymerization to decrease with increasing X, contrary to what experiments consistently show (see, e.g., Refs. [51,96]). In section 5, a simple model will be presented that avoids these shortcomings and is able to correctly predict the chain length data of Albouy et al. [79].

# 4 Energetics of the Polymerization Process

Although kinetically hindered by an energy barrier attachment of a monomer to a running polydiacetylene chain is an exothermic process. This has been proven by Chance et al. [97] employing differential scanning calorimetry. These authors recorded the heat evolved during thermal reaction of TS and found that an energy of 1.6 eV is released by adding one monomer to an active chain end. Experimentally, a TS-sample is held at constant temperature in the calorimeter. Onset of the reaction after the induction time shows up as a peak in the DSC scan. Since conversion is quantitative, this experiment directly yields the reaction enthalpy per monomer addition event.

Later on, Chance and Shand [98] studied the heat evolution during UV-photopolymerization of 4-BCMU employing the photoacoustic detection technique. This technique is based on the increase of the pressure of the gas surrounding with a sample in which heat is produced as a result of light absorption. Although highly sensitive, quantitative data analysis is not straightforward since it requires taking into account both heat diffusion inside the sample and the magnitude of the acoustic coupling between solid and gas [73]. The information it provides is the reaction enthalpy produced per chain initiation event, i.e., $nq\Delta H$. It therefore can be used to determine $nq$ once $\Delta H$ is known from DSC studies. Being a transient technique, PA-photocalorimetry can also monitor the time response of the sample towards a light pulse that triggers the photoreaction. For 4-BCMU it was found that chain formation must be completed within a time $<1\ \mu s$, which was the time response of the detector system.

More direct information on the energetics of the photopolymerization process is afforded by DSC measurement with UV-initiation of the reaction. Under single pulse excitation the integral heat evolved per light pulse is a measure for the product $n\Delta H$ and from the ratio of the signal amplitudes at $t = 0$ and $t \rightarrow \infty$, respectively, the quantity nqΔH can be inferred independently. Thus, if the chain length n is known from measurements of the molecular weight, as it is the case with soluble diacetylenes of the BCMU-type [84], both ΔH and q can be determined. Employing this technique, Eckhardt et al. [74] recently showed that in 4-BCMU, $\Delta H = 0.95$ eV per repeat unit, i.e., about 0.6 eV less than in TS. They proposed an explanation involving the intramolecular hydrogen bonds between the C=O and N—H functionalities on adjacent substituents of diacetylene monomers, known to exist both in solution and in the solid state [85]. Based on steric arguments it has been concluded that in 4-BCMU, where four $CH_2$ groups link the side groups to the backbone, a considerable distortion of the $CH_2$-groups and possibly also the backbone is required for preservation of the hydrogen bonds after polymerization. Therefore, polymerization goes in hand with an energetically unfavorable reorganization of the intramolecular hydrogen bond network at the expense of the total reaction enthalpy. This reasoning provides a straightforward explanation why 4-BCMU does not polymerize thermally at any measurable rate. Formation of the chain initiating dimer is an endothermic process and involves similar rearrangement in the side groups as does addition of a monomer to a running chain. Therefore, the extra energy of ~0.6 eV resulting from H-bond distortion should be added to the energy barrier for chain initiation, resulting in a drastic decrease of the thermal reaction rate.

The essence of the energetic studies on TS and 4-BCMU is contained in Fig. 9. In TS formation of the chain initiating species — a dimer — requires an energy of 1.0 eV. It can be supplied thermally or optically via monomer excitation. In the former case it is this chain initiation reaction that controls the thermal reactivity and its temperature-dependence. Chain initiation can also be produced optically at a yield of order $10^{-2}$ per absorbed UV-quantum. In this case it is chain propagation that determines the temperature dependence of the polymerization yield. However, the activation energy $E_a^{opt}$ need not be and in general is not identical with the energy

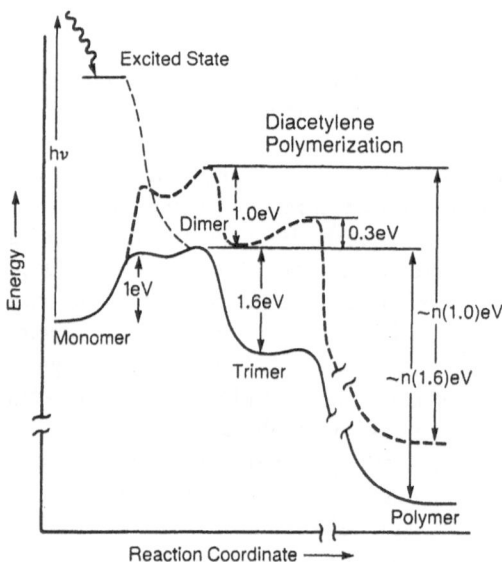

**Fig. 9.** Energy diagram for polymerization of diacetylenes. The solid curve is for TS-6 (Ref. [98]), the dashed curve for 4-BMCU (Ref. [74])

barrier for monomer addition. It is a measure for the temperature dependence of nq, or, rather, of n, since there is good reason to suspect that except at low temperatures q shows only a weak if any temperature dependence. In case that the chain length is not saturated, i.e., limited by the number of monomers available for reaction, n is given by $k\tau$ where k is the rate constant of the addition reaction and $\tau$ the lifetime of the active chain end. Even then $E_a^{opt}$ is not identical with the height of the kinetic barrier since $\tau$ is also temperature dependent. Comparing $E_a^{opt}$-data with the results of kinetic studies [67] (see Sect. 5) in fact indicate, that in TS the true barrier height for the monomer addition reaction is $> E_a^{opt}$ and dependent on the length of the precursor chain.

Eckhardt et al. [74] also speculated on the possible reason for the high photochemical yield of 4-BCMU as compared to TS. Even taking into account the much longer chain length in 4-BCMU, the chain initiation probability q is still coming out about a factor of 4 larger than in TS. These authors argued that because of the upward shift of the energy of the chain initiating species (see above) its electronic level comes closer to the electronic state of the excited monomer favoring non-radiative coupling between those states. This suggests existence of an inverse relationship, i.e., a compensation effect between optical and thermal reactivity of diacetylenes. Clearly, more quantitative work is needed to test this interesting hypothesis.

There has been an argument in the literature whether the intermediate acting as a chain initiator is a dicarbene of the type:

$$\cdot\overset{\displaystyle R}{\underset{\displaystyle R}{C}}-C\equiv C-\overset{\displaystyle R}{\underset{\displaystyle R}{C}}\!\!\left(\!\!\overset{\displaystyle R}{\underset{\displaystyle R}{C}}-C\equiv C-\overset{\displaystyle R}{\underset{\displaystyle R}{C}}\!\!\right)_{n-2}\!\!\overset{\displaystyle R}{\underset{\displaystyle R}{C}}-C\equiv C-\overset{\displaystyle R}{\underset{\displaystyle R}{C}}\cdot$$

or a diradical [53]:

$$\cdot\overset{\displaystyle R}{\underset{\displaystyle R}{C}}=C=C=\overset{\displaystyle R}{\underset{\displaystyle R}{C}}\!\!\left(\!\!\overset{\displaystyle R}{\underset{\displaystyle R}{C}}=C=C=\overset{\displaystyle R}{\underset{\displaystyle R}{C}}\!\!\right)_{n-2}\!\!\overset{\displaystyle R}{\underset{\displaystyle R}{C}}=C=C=\overset{\displaystyle R}{\underset{\displaystyle R}{C}}\cdot$$

The argument in favour of the latter was that the diradical should be lower in energy, because its formation requires disruption of only one carbon-carbon π-bond instead of two in case of carbene formation. Both ESR work [22] and optical spectroscopy [17] have meanwhile confirmed the diradical mechanism for growth of oligomeric chains up to length of 5 repeat units. Upon further addition of monomers, the acetylenic structure becomes energetically more stable causing a cross-over to the carbene mechanism. For further discussion of this topic the reader is referred to the article by H. Sixl in this volume.

## 5 Time Dependent Studies

### 5.1 Studies at Atmospheric Pressure

A major achievement towards understanding the kinetics of photopolymerization of diacetylenes was the application of time dependent optical spectroscopy. Monitoring the appearance of the characteristic polymer reflection Leyrer and Wegner [100] were the first to record the time-dependent build-up of polymer in TS-6 and MCD following flash excitation. The light source was a $N_2$-laser delivering 5 ns pulses which drive the lowest forbidden singlet transition (at 3.7 eV) of the diacetylene moiety [50]. Polycrystalline powder dispersed on filter paper was used as a sample. Although in a powder sample the density of structural defects which might conceivably block chain growth, inevitably is high, the experiment revealed the essential features of polymer formation correctly. Fig. 10, which is taken from Ref. [100], shows that

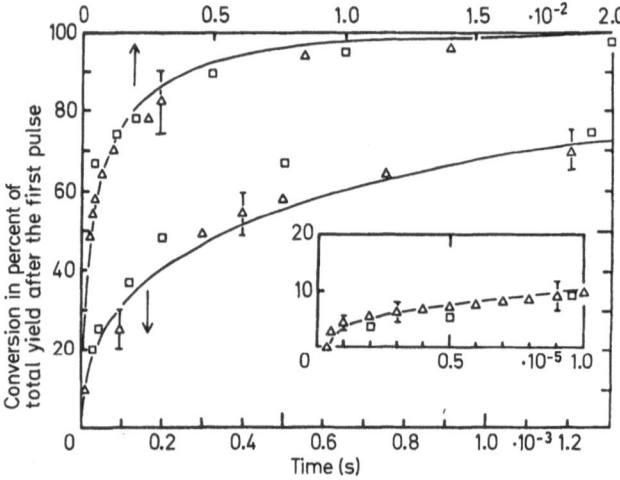

**Fig. 10.** Time dependent conversion of TS-6 (squares) and MCD (triangles) following one $N_2$-laser excitation pulse of 3 ns duration. The inset shows the MCD-data on an enlayed scale. Data are normalized to the polymer content in TS-6 reached in the $t \rightarrow \infty$ limit (from Ref. [100])

build-up of the polymer is a relatively slow process, occurring on a time scale of the order 1 ms, and does not follow simple first order kinetics. Empirically a $X(t) \sim$
$\sim 1 - \exp\left[-(C(t - t_d))^{1/2}\right]$ law was obeyed with $C = 1.3 \cdot 10^3 \text{ s}^{-1}$ and a delay time $t_d = 300$ ns elapsing between laser flash and onset of polymer formation. The $\sqrt{t}$-law led Leyrer and Wegner to conclude that it is the diffusion of excited states within the monomer matrix towards reaction centers, possibly defects that determines the kinetics. Subsequent time resolved absorption studies by Niederwald, Eichele, and Schwoerer [101] performed on single crystals did not confirm this aspect of Leyrer and Wegner's interpretation but showed that the observed time dependence is the necessary consequence of the multistep character of the chain propagation process. Performing either a monochromatic or a polychromatic transient absorption experi-

ment, Niederwald and Schwoerer [67] recently succeeded in recording both absorption spectra and transient behavior of short-lived reaction intermediates appearing in a TS crystal after excitation by a 15 ns laser pulse of $\lambda = 308$ nm and a pulse energy of typical 1 mJ. Figure 11 presents the transient difference absorption spectra at

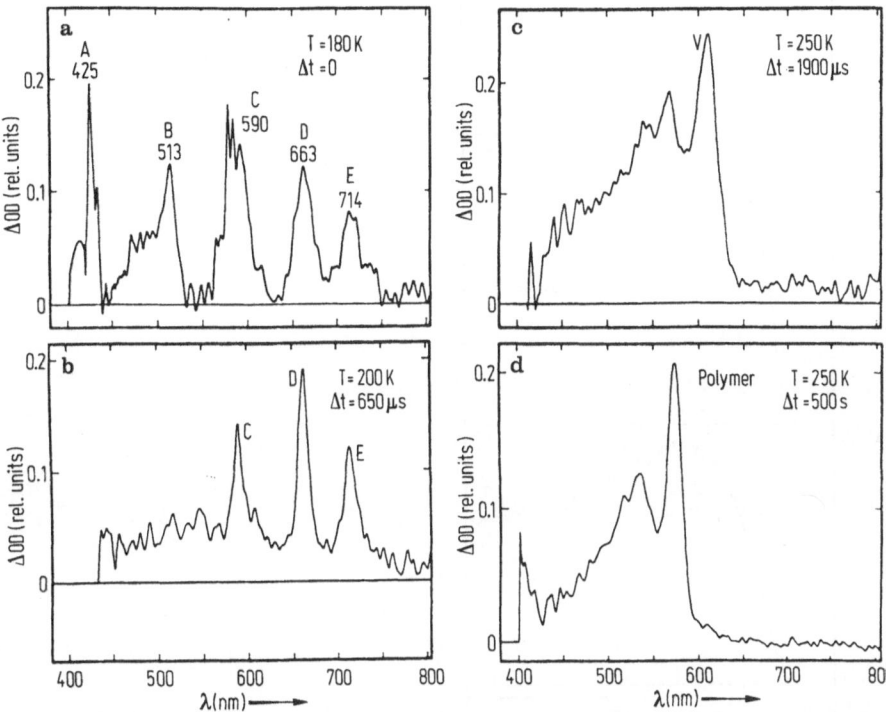

**Fig. 11a–d.** Transient absorption spectra of reaction intermediates (a–c) and final product (d) observed upon photopolymerization of a TS-6 crystal with a 308 nm laser pulse. Parameters are the temperatures and the delay times ($\Delta t$) between the laser pulse and spectra recording. In b and c the time window of the detection circuit was 100 $\mu$s. $\Delta OD$ is the difference of the optical densities after and before the UV-flash (from Ref. [67])

180 K. Spectrum $a$ was recorded within 100 $\mu$s after the UV flash, spectra $b$ and $c$ were recorded within a time window of 100 $\mu$s after a delay time of 650 $\mu$s and 1.9 ms, respectively. The conclusion is, that a series of five intermediates A, B, C, D, E exists, whose lifetime increases with increasing absorption wavelength. The absorption spectrum of product V closely resembles the final polymer spectrum (11d). Following Sixl et al. [14, 15] these intermediates are identified as dimer (A), trimer (B), tetramer (C), pentamer (D), and hexamer (E)-diradical ($DR_n$).

Growth and disappearance of the intermediates, monitored by their characteristic transient absorption (Fig. 12) can be fitted perfectly on the basis of the kinetic scheme:

$$A \xrightarrow{K_A, \Delta E_A} B \xrightarrow{K_B, \Delta E_B} C \xrightarrow{K_C, \Delta E_C} D \xrightarrow{K_D, \Delta E_D} E$$

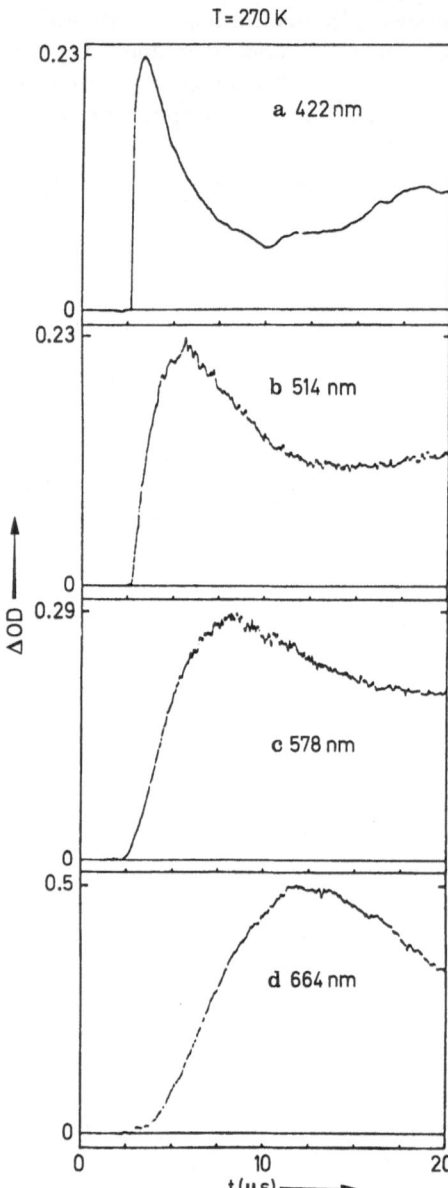

T = 270 K

a 422 nm

b 514 nm

c 578 nm

d 664 nm

**Fig. 12.** Time dependent absorption of reaction intermediates A to D observed upon UV-photopolymerization of TS-6 crystals at 270 K. ΔOD is the change in optical density of the sample (from Ref. [67])

where the K's are rate constants in first order rate equations and the ΔE's are activation energies. Solution of the pertinent set of coupled differential equations shows that the concentration of the i-th product is:

$$n_i(t) = \sum_{j=1}^{i} a_{ij} \exp\left[-K_j(t - t_0)\right] \tag{6}$$

where $a_{ij} = a_{i-1,j}\dfrac{K_{i-1}}{K_i - K_j}$, $i, j \hat{=} A, B, C, D, \ldots$, and $t_0$ is the time at which the

precursor to A, assumed to be the photoexcited monomer, is populated. For data analysis it is further assumed that population of A occurs fast on the time scale of the transient absorption experiment. Verification of the above simple reaction scheme confirms the earlier notion that each reaction step consists of the addition of one monomer unit to a linear chain. As the length of t·e oligomers increases beyond a certain limit $(i > 5)$ [17] their diradical character changes to dicarbenic or carbenic and the absorption bands gradually converge. Product V is interpreted as a superposition of long chain oligomers which absorb at a wavelength close to the value characteristic of the infinite chain. Kinetically, V is therefore regarded to be the ultimate precursor for the final polymer. It is gratifying to note that the time dependence of polymer growth measured by Leyrer and Wegner agrees with single crystal data, documenting that reaction kinetics — at least towards the end of chain growth — is fairly insensitive to changes in the macroscopic perfectness of the sample. Apparently, in the low conversion limit the kinetic chain length in TS-6 is much less than both the dimension of crystallites and the mean distance between lattice defects which could block chain growth.

Interestingly, the preexponential factor of the rate constants is practically the same for all intermediates $(K_0 = 10^{11 \pm 1}\ s^{-1})$, whereas the activation energy slightly increases from $\Delta E_A = 0.25 \pm 0.03\ eV$ to $\Delta E_{V2} = 0.40 \pm 0.03\ eV$ (see Fig. 13).

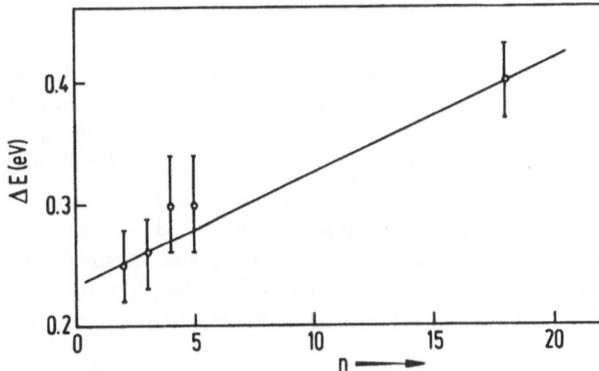

**Fig. 13.** Activation energy for the addition reaction n-oligo-mer + monomer → (n + 1) oligomer as a function of the number of repeat units (n) of the oligomer (data are taken from Ref. [67])

This suggests that the addition reaction is a thermally activated process, driven by a lattice mode with frequency of the order $10^{11}\ s^{-1}$ corresponding to an energy of about $30\ cm^{-1}$, typical for optical or acoustic lattice modes. Remarkably, Niederwald and Schwoerer [67] quote recent neutron diffraction work by Schott and coworkers delineating existence of an optical phonon at a very low frequency ($\sim 10^{11}\ s^{-1}$) at the boundary of the Brillouin zone.

Qualitatively, the increase in activation energy with increasing length of the oligomer can be attributed to the mismatch between monomer lattice and length of the oligomer/polymer repeat unit [52]. As the oligomer length increases, the gap between the oligomer head and the adjacent monomer widens. Therefore, more

thermal energy is required to establish the critical distance necessary for bond formation between the adjacent molecules. This idea will be elaborated further in Section 6
taking advantage of kinetic studies under hydrostatic pressure to be discussed in
section 5.2.

An interesting isotope effect on both thermal- and photopolymerization of specific-
ally deuterated or $^{13}$C-labelled TS-6 was discovered by Kröhnke, Enkelmann, and
Wegner [102]. These authors found that deuteration in the $CH_2$-group and/or the
phenylring of the toluenesulfonate side group reduces both induction time for thermal
polymerization and reaction rate within the autocatalytic regime by typically a factor
of 2. The largest effect was seen upon isotopic substitution at the methylene group
which forms the linkage between diacetylene moiety and substituent. Verification
of this effect with HDU, a phenylurethane-substituted diacetylene, suggests that it
is a general feature of the polymerization of diacetylenes rather than an impurity
effect [103]. It reflects the influence of isotopic substitution on both lifetime of the chain
propagating species and the rate constant for the monomer addition reaction [67, 102].
The lifetime $\tau$ is determined by the decay of a reactive chain end of the carbene type
to a stable polymer head involving electronic rearrangement and singlet-triplet
intersystem crossing, known to become less efficient upon deuteration [104]. An
increase of $\tau$ allows polymer chains to grow longer. In course of a thermal polymerization experiment the critical conversion above which autocatalytic reaction enhancement becomes effective is therefore reached at shorter times in the deuterated material
provided the chain initiation rate remains constant, i.e. the induction time decreases.
On the other hand, the reaction rate within the autocatalytic reaction regime decreases
reflecting the decreased rate of the monomer addition reaction in the deuterated
material. This is in accord with the experiments of Kröhnke et al. [102].

A final remark regarding the maximum kinetic chain length in the absence of any
other factors limiting chain growth appears in order. In this case the rate for monomer
addition should be constant throughout the chain growing process and the chain
length should simply be $n = k\tau$ repeat units. Assuming $k \approx k_A = 10^6 \text{ s}^{-1}$ and
$\tau \approx 10^{-3}$ s gives $n \approx 1000$ as an order of magnitude estimate. It agrees with the
average chain length measured with TS-12 [86] as well as 3-BCMU and 4-BCMU [84].

On the other hand, anticipation of a lifetime in the order 1 ms is at variance with
the observation made in course of photoacoustic experiments that the photoreaction
is accomplished within $10^{-6}$ s [98]. Clearly, additional time resolved studies are
required to clarify this problem.

## 5.2 Pressure Studies

The motivation for conducting high pressure studies on the polymerization of TS
was the desire to decide whether or not the probability for initiating a polymer chain
depends on the distance of the monomers forming the dimer. Remember that the
elastic strain theory predicts strain to affect both chain propagation and chain initiation
in the same manner. If reduction of the average lattice distance along $\vec{b}$ accompanying
the autocatalytic reaction enhancement in TS were to increase the chain initiation
rate, application of hydrostatic pressure should be similarly effective. In case that

chain initiation is accomplished by photons, the relative number of chain-initiators is amenable to experimental probing by time resolved absorption spectroscopy.

The experiments were similar to those of Niederwald and Schwoerer [67], except that the crystal was mounted in a high pressure cell operated with water as pressure transmitting medium and that measurements were done at ambient temperature only [72]. Figure 14 shows a series of absorption transients associated with intermediate B

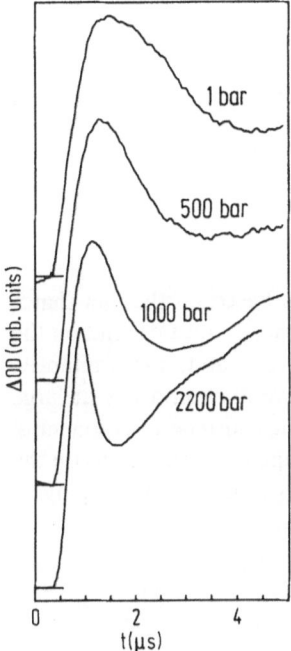

**Fig. 14.** Time dependent absorption of the reaction-intermediate B (trimer) observed upon UV-polymerization of TS-6 under hydrostatic pressure (a: 1 bar, b: 0.5 kbar, c: 1.0 kbar, d: 2.2 kbar). Excitation was by a single flash of excimer laser operated at 308 nm (from Ref. [72])

(see Sect. 5.1), which is the diradical trimer, recorded at variable pressure after a single UV-pulse. Each curve was taken with a new crystal. It is obvious, that both rise and decay of intermediate B become faster with increasing pressure. This implies that within the accuracy of the experiment the rate constants $K_A$ and $K_B$ are similarly affected by pressure and their relative variation can be infered from the pressure shift of the peak of the transient absorption signal. Figure 15 reveals an exponential increase of K with pressure, $K(p) = K_0 \exp (p/p_0)$ with $p_0 = 2.8 \pm 0.2$ kbar. Despite a large scatter in absolute values, inevitably caused by changing samples in each experiment, no pressure induced increase of the signal amplitude, as a measure of the number of trimeric diradicals, is noted. This provides experimental proof that pressure does not noticeably affect chain initiation but does aid chain propagation. By analogy, one has to conclude that the rate of thermal chain initiation is also independent of internal pressure, generated, for instance, by chains already present within a polymerizing matrix. This casts doubts on the applicability of the elastic strain theory to analyze polymerization kinetics of diacetylenes quantitatively.

In the following chapter a model will be presented that relates the velocity of the

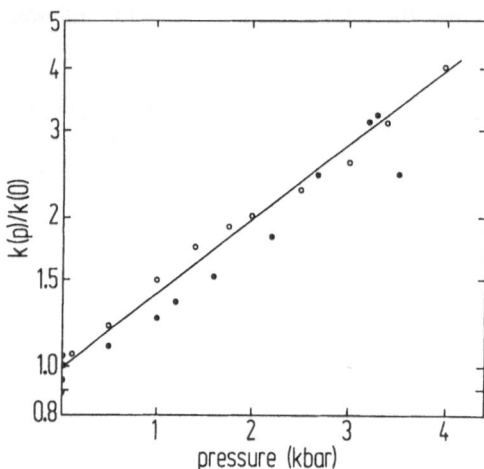

**Fig. 15.** Pressure dependence of the normalized rate constant for formation and decay of reaction intermediate B. Open and full circles refer to different series of experiments (from Ref. [72])

reaction to the distance between the reacting centers. It is based on the fact that a chemical reaction, in course of which a covalent bond is formed, must depend on the overlap of the electronic wavefunction of the reacting species and, concomitantly, on their mutual distance s. Since in organic molecules wavefunction decay at large distances from the nuclei is exponential, the overlap factor must also be an exponential function of s. It will be shown that the effect of structure dependent variations on the overlap factor overrides the direct effect of the strain energy changes developing as a result of lattice mismatch between monomer and polymer.

## 6 Model Considerations

### 6.1 Chain Propagation

In a TS-6 monomer crystal the distance s between the C1 and C4 carbon atoms of adjacent molecules is close to 3.6 Å. After formation of an oligomer s is reduced to approximately 2.5 Å (see Fig. 16), still too large to allow instantaneous reaction as evidenced by appearance of an energy barrier of typically 0.25 eV controlling the reaction n-oligomer + monomer → (n + 1)-oligomer. It is straightforward to associate this activation energy with the excitation of librational motion of the diacetylene moiety. In its course the reaction distance is temporarily reduced to a value $s_m$ characteristic of the maximum in an energy versus reaction coordinate-diagram.

The barrier height for the reaction to proceed can be expanded into a power series on $s - s_0$. Truncation at the linear term gives $E = \dfrac{\delta \Delta E}{\delta s}(s - s_0)$. Support for the applicability of the linear approximation comes from the data of Ref. [67]. They show that the activation energy for adding a monomer to the i-th oligomer rises linearly with i. Conversion of intermediate V into the final polymer product requires an activation energy $\Delta E_V = 0.4 \pm 0.03$ eV. Since this reaction shows a more complicated

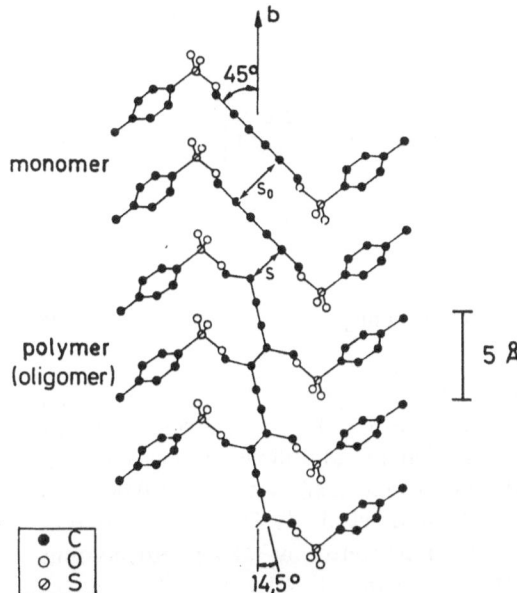

monomer

polymer
(oligomer)

5 Å

Fig. 16. Atomic positions in the monomer and polymer (oligomer) of TS-6 as deduced from X-ray data (Ref. [92]). $s_0$ and $s$ are the C1–C4 separations (reaction distances) for addition of a monomer to another monomer and an oligomer, respectively

time behavior, it cannot a priori be identified with monomer addition to an oligomer having a length characteristic of TS-6 in the X → 0 limit, which is about 18 repeat units (see above). Therefore the above value represents an upper limit for the activation energy governing the addition reaction before chain termination. Since, on the other hand, the ΔE-data obtained for short oligomers tend to extrapolate to 0.4 eV at n = 18, one can to first order approximation identify this value with the final monomer addition step. It follows that in the low conversion limit the increase of the activation energy $\delta\Delta E$ per addition reaction is $\left(\dfrac{\delta\Delta E}{\delta n}\right)_{X=0} = (9 \pm 2) \cdot 10^{-3}\,\text{eV}$ (see Fig. 13).

The increase of ΔE with chain length must originate from the difference between the monomer lattice constant and the length of oligomer/polymer repeat unit. Anticipating proportionality between $\delta\Delta E/\delta n$ and lattice mismatch as suggested by studies under pressure (see 5.2), the variation of ΔE with n can be calculated for arbitrary conversion X from the low conversion value $(\delta\Delta E/\delta n)_{X=0}$:

$$\frac{\delta\Delta E(X)}{\delta n} = \left(\frac{\delta\Delta E}{\delta n}\right)_{X=0} \frac{b(X) - b_p}{b_m - b_p} \qquad (7)$$

Here $b_m$ and $b_p$ are the lattice constants of monomer along $\vec{b}$ and the length of the oligomer/polymer repeat unit, respectively, and b(X) is the average b-axis dimension at arbitrary conversion X.

Eq. (7) can be used to calculate the number of repeat units n contained in the polymer as a function of conversion. In the low conversion range, n is given by the number of addition steps occurring per lifetime τ of the reacting chain end, $\tau = \sum\limits_{i=1}^{n} k_i^{-1}$.

The rate constant for the i-th reaction step, $k_i$, depends upon i according to:

$$k_i = k_0 \exp - \left[ \frac{\Delta E_1 + \frac{\delta \Delta E}{\delta n} i}{kT} \right] = k_1 \left[ \exp - \frac{\delta \Delta E}{\delta n kT} i \right] \qquad (8)$$

After replacing i by the continuous variable n and converting $\sum_{i=1}^{n} k_i^{-1}$ into an integral, Eq. (8) yields:

$$n(X) \cong \left[ \frac{\delta \Delta E(X)}{kT \, \delta n} \right]^{-1} \ln \left[ \tau k_1 \, \delta \Delta E(X)/kT \delta n \right] \qquad (9)$$

Inserting experimental data for $(b(X) - b_p)/(b_m - b_p)$ from Fig. 7, $(\delta \Delta E/\delta n)_{X=0}$ = $9 \cdot 10^{-3}$ eV (see above), $k_1 = 10^6$ s$^{-1}$, and $\tau = 7.3 \cdot 10^{-4}$ s at 295 K [67] allows calculating absolute numbers for n(X). Note that no adjustable parameter enters. Figure 4, which shows the result of the calculation, demonstrates that Eq. (9) provides an excellent fit to the experimental data of Albouy et al. [79]. This documents that chain propagation in TS can be consistently interpreted by taking into account the variation of the reaction rate constant with reaction distance.

Unfortunately, the above data relating activation energy $\Delta E_i$ to the number of addition events are insufficient for constructing the energy profile of the reaction as a function of reaction distance s. This is because both oligomers and polymers formed under low conversion conditions are not in their relaxed state but are expanded as evidenced by their blue-shifted absorption. Although an average value for the lattice parameter along $\vec{b}$ will be established there will be mismatch on a molecular level accounting for the increase of $\Delta E$ discussed above. Therefore, absolute reaction distances are not available at a level of accuracy required to render an estimate of the energy profile meaningful. The missing information can, however, be extracted from the pressure effect on the reaction constant.

As mentioned above, the rate constant for generation and decay of intermediate B varies under pressure as $k(\Delta p) = k_0 \exp (\Delta p/p_0)$ with $p_0 = 2.8 \pm 0.2$ kbar at 295 K. Since the reaction is thermally activated, the pressure effect must be due to a reduction of the activation energy. Comparing exponents gives:

$$\Delta p/p_0 = -\delta \Delta E/kT \qquad (10)$$

On the other hand, the pressure-induced change in the reaction distance s is:

$$\Delta s = s \chi_s \Delta p \qquad (11)$$

where $\chi_s$ is the compressibility along the lattice vector joining the C1 and C4 positions of the reacting molecules. A reasonable estimate of $\chi_s$ can be infered from the linear compressibility $\chi_b$ of the monomer lattice, known to be $\chi_b = 4.2 \cdot 10^{-3}$ kbar$^{-1}$ [96]. Because in the monomer lattice the diacetylene moiety is at an angle of 45° with respect to $\vec{b}$ (see Fig. 16) the torque felt by the $-C \equiv C - C \equiv C -$ unit upon applying hydrostatic pressure is likely to cancel out and the lattice compression along $\vec{b}$ will fully be converted into a reduction of the molecular center of mass distances. From geometrical consideration $\chi_s = 6 \cdot 10^{-3}$ kbar$^{-1}$ and, since $s \approx 2.5$ Å, $\Delta s = 1.5 \times 10^{-2}$ kbar$^{-1}$ $\Delta p$ Å follows. Inserting this value into Eqs. (10) and (11) predicts $\delta\Delta E/\delta s = 0.6$ eV/Å for the slope of the energy barrier in the range where it is linear with reaction distance. This information allows to estimate the location of the maximum of the potential barrier in an E(s)-diagram. To attach a monomer to intermediate B requires an activation energy $\Delta E_B = 0.25$ eV at an initial reaction distance $s_B \approx 2.5$ Å. To reduce $\Delta E_B$ to zero requires a reduction of s by 0.4 Å to $s_m \sim 2.1$ Å. In order to be able to react with the active end of an oligomer an adjacent monomer has therefore to execute a thermally activated torsional motion with an amplitude allowing approach of the C1 and C4 carbon atoms to $s_m \lesssim 2.1$ Å.

To a first order approximation the energy profile of the reaction must be the superposition of the vibrational energy of the monomer as a function of vibration amplitude and the electronic binding energy of the reaction product as a function of bond-distances. To be realistic it must be in accord with the following experimental data: (i) $E(s = s_m) - E(s_0 = 3.6$ Å$) \leq 1.0$ eV which is the energy required to initiate a chain, (ii) $E(s = s_m) - E(s = 2.5$ Å$) = 0.25$ eV, (iii) $\delta\Delta E/\delta s \approx 0.6$ eV/Å, and (iv) $E(s \sim 2.5$ Å$) - E(s = 1.4$ Å$) = 1.6$ eV which is the amount of energy released upon addition of one monomer unit. Figure 17 shows an energy diagram which matches the above conditions. It is constructed from a vibrational potential in harmonic approximation $V_V(s - s_0) = V_0 + \frac{1}{2}k(s - s_0)^2$ with $s_0 = 3.6$ Å and a

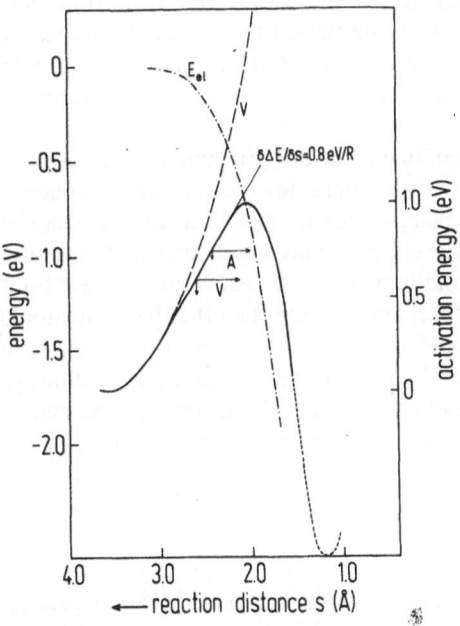

**Fig. 17.** Energy profile for polymerization of diacetylenes. Numerical data are for TS-6. $----$: harmonic potential (V) for librational motion of the diacetylene moiety. $-\cdot-\cdot-\cdot-$: electronic binding energy ($E_{el}$); the full curve is the sum of $E_{el}$ and V. The origin of the energy scale (left ordinate) is chosen so that $E_{el}(s \to \infty) = 0$. The right ordinate gives the activation energy required for a monomer addition reaction as a function of reaction distance s. A and V stand for reaction intermediate A (dimer) and V (precursor to stable polymer), respectively

force constant $k = 0.72$ eV/Å equivalent to 23 N/m and an electronic binding energy $E_B = E_B(s \to \infty) - \beta \exp(\alpha s)$ with the parameters $\beta = 177$ eV and $\alpha = 2.68$ Å$^{-1}$. The same value for $\alpha$ has been used previously to approximate the distance dependence of the overlap integral for carbon-carbon bonds in diacetylenes [105]. The force constant turns out to be about two orders of magnitude smaller than force constants characteristic for double bond stretching modes. This is in accord with the finding that it is a low frequency ($\approx 30$ cm$^{-1}$) librational motion of the diacetylene moiety which drives the reaction [67]. The E(s)-profile exhibits a linear regime over a wide range of s-values explaining why (i) the thermal activation energy of the addition reaction is a linear function of the relative lattice mismatch and (ii) the rate constant varies exponentially with pressure. Contrary to conventional energy diagrams where the potential energy is plotted versus a generalized reaction coordinate, E(s) gives the energy required by a monomer molecule to react with one of its neighbors as a function of only the C1–C4 reaction distance.

The above treatment for calculating n(X) is readily extended to cover the whole conversion range. To facilitate comparison with the prediction of the elastic strain theory the result of the calculation has been included in Fig. 8. It is obvious that n(X) rises more slowly with increasing conversion than predicted by the ES-theory. Following the principle that it is the slowest process that dominates a reaction, it is concluded, that growth of a polydiacetylene chain is essentially governed by the reaction barrier resulting from a superposition of the vibrational potential and the electronic overlap of the reacting molecules. The effect of the elastic strain is to establish the equilibrium reaction distance entering the expression for the activation energy. It does not appear to control chain growth in a direct way. This explains why often difficulties are encountered when reaction kinetics are analyzed in terms of the ES-theory [95].

At this stage it seems appropriate to comment on the principle of least motion introduced by Baughman [9] in order to establish criteria for the reactivity of diacetylenes. It says that a reaction will only take place if the root mean square displacement of the atoms does not exceed a certain critical value. This means that each displacement is counted at equal weight. The dominant effect a reduction of the C1–C4 distance has on the reaction rate is, therefore, ignored. On the other hand, it should be pointed out that the treatment advanced above, which relies on the dominant role of the C1–C4 reaction distance, does embody all atomic displacements, although in an implicit way via the vibrational energy which essentially determines the energy profile for the reaction, and which depends on the structure of the reactants.

## 6.2 Chain Initiation

The energy diagram presented in Fig. 17 is of relevance for the chain initiation process as well. It demonstrates that the 1 eV barrier for chain initiation is determined by the activation energy of the reaction rather than the energy difference between dimer and monomer. Since the reaction requires reduction of the reaction distance from 3.6 Å to ~2 Å, both reacting monomers have to execute torsional motions. Otherwise the diacetylene moiety of one monomer had to be rotated by about 45° into a position parallel to the crystallographic b-axis. This is incompatible with steric constraints and explains the observation of Neumann and Sixl [16] that at low temperature, where librational motion is frozen-in, photoinitiation is a two quantum process. In this case the motional energy of both monomers has to be supplied externally. Further, the parabolic shape of the potential profile near the monomer equilibrium distance indicates that lattice compression either by applying hydrostatic pressure or by polymer chains already formed, will not reduce the activation energy to any measurable extent. This resolves the paradox encountered when applying the elastic strain theory to explain both autocatalytic and pressure-induced reaction acceleration. Remember that this theory predicts a lowering of the energy barrier both for chain propagation *and* chain initiation, quite in contrast to all experimental reports.

It is tempting to speculate on the possibility that optical and thermal chain initiation occur via basically the same mechanism, the only difference being the way of accumulating the activation energy at the reacting pair of molecules. However, this hypothesis is readily discarded by comparing the preexponential factors for thermal chain initiation and chain propagation, respectively. Whereas, for the latter process $k_0 = = 10^{11 \pm 1}$ s$^{-1}$ has been reported [67], it is about four orders of magnitude lower for the former [97]. Since in both cases the excess energy is supplied thermally, no significant changes in the partition function entering the equilibrium constant of the transition state are expected to occur. It therefore appears that once it has reached the top of the potential barrier a pair of monomers has a $10^4$-fold higher chance to react if one of the pair partners has been optically excited. The most likely explanation is that either orbital symmetry or spatial extent of the wavefunction are more favorable for reaction while the monomer is in an excited state.

There has been discussion in the literature [53] as to whether the optically excited monomer precursor state for the dimer is of singlet or triplet character. Consider that the diacetylene $S_{1D}$-state undergoes intersystem crossing to the triplet manifold at a yield $q_{ISC}$. Following Chance and Patel [53] the total probability for chain initiation after excitation of the $S_{1D}$-state by an UV-quantum can then be written as the sum of a singlet and a triplet term, $q^{UV} = q^S + q_{ISC}q^T$ where $q^S$ and $q^T$ are the chain initiation probabilities for a monomer in its singlet and triplet state, respectively. If, on the other hand, the reaction is initiated by $\gamma$-irradiation, the primary yield, expressed in terms of the G value (number of products per 100 eV absorbed energy), is $G = G^S q^S + G^T q^T$ where $G^S$ and $G^T$ are the primary singlet and triplet yields, respectively. From the organic scintillator literature it is known [106] that $G^S \approx G^T \approx 1$. Experimentally one finds [53], that in TS-6 about 40 monomers are polymerized per 100 eV absorbed energy, i.e., $G \approx 2$ assuming a chain length of 20 repeat units. In TS-12 the G-value for chain initiation is 1.2 [88]. It thus follows that $q^S + q^T \approx 1 \ldots 2$. Since $q^S < q^{UV} \approx 0.005$, $q^T$ must be of order unity. This estimate lends support for assigning triplet character to the optical precursor state for chain initiation and for claiming that dimer formation is the dominant channel for decay of the monomer triplet state.

Albeit direct experimental evidence for polymerization upon direct optical excitation into the triplet manifold of a diacetylene is still lacking several arguments support the above reasoning.

(i) Steric arguments require that electronic excitation of the diacetylene moiety has to be accompanied by vibrational excitation in order to promote dimer formation. Otherwise the C1–C4 distance would be prohibitively large. This condition is met if the active state is populated via a nonradiative transition from a higher excited state. UV-photopolymerization of TS is normally done by $h\nu > 4$ eV excitation. Since the lowest triplet state of diacetylene is located at 3.1 eV [107], about 1 eV of excess energy is available for conversion into vibrational exitation. The above estimate requires that the yield for populating the active triplet state by intersystem crossing be $q_{ISC}^{eff} \sim 4 \cdot 10^{-3}$, considerably less than expected for $S_1 \to T$ conversion across an energy gap of $<1$ eV. From spectroscopic work it follows that in this case $q_{ISC} \geq 0.1$ [108]. The ratio $q_{ISC}^{eff}/q_{ISC} \sim 0.04$ could be interpreted as the probability for funnelling the excess energy into the active librational modes. An independent indication for the importance of vibrational energy in addition to electronic excitation is the increase of the quantum yield nq with photon energy (see Fig. 3 in Sect. 2).

(ii) ESR work indicated that the carbene-type as well as the diradical intermediates have triplet character. The long lifetime $\tau$ of the active chain end, typically 1 ms, is likely to originate from a slow triplet-singlet intersystem crossing process. Since all rate constants for formation of short oligomeric species including the diradical dimer are $\gg \tau^{-1}$, typically $10^6 s^{-1}$, intersystem crossing processes cannot be involved in these processes. This suggests that the optical precursor state for the dimer has already triplet character. In fact, Gross et al. [17] demonstrated that kinetic data for TS-6 can consistently be interpreted if the diradical dimer is assumed to be produced via a vibrationally hot triplet state of the monomer. On the basis of time-resolved studies its lifetime is estimated to be 50 to 100 ns [101] and is likely to be reaction-limited (see above).

(iii) The small preexponential factor for thermal chain initiation, could reflect the low probability for $S \to T$ conversion in course of dimer formation involving thermal excitation of a pair of monomers in their singlet ground state.

In this context it seems noteworthy that nq is about four orders of magnitude smaller for DCH [109] as compared to TS-6. Since contrary to TS, in DCH both singlet and triplet levels of the carbazole substituent ($E_{S1} = 3.4$ eV, $E_{T1} = 2.9$ eV [110]) are below the corresponding levels of the diacetylene linkage, excitations of the latter are efficiently quenched. Therefore, UV-photopolymerization of DCH must in fact occur at a much lower yield, independent of whether the precursor state is of singlet or triplet character. Surprisingly, the G-value for γ-polymerization is only about one order of magnitude less as compared to TS [109]. This could be an indication that, in addition to the triplet route, charge carriers, which are produced by γ-irradiation yet not by UV-excitations, also act as chain initiating species. Support for this notion comes from the experiments on self-sensitized and sensitized photopolymerization to be discussed in Section 8, which also draw heavily on the role of charge carriers.

In summary, it is well established that the species that initiates exothermic growth of a polydiacetylene chain is a diradical dimer. It can be generated thermally, the activation energy being determined by the energy of the librational motion required to temporarily shorten the C1–C4 reaction distance of a molecular pair to about 2 Å, or by electronic excitation of the diacetylene moiety via UV- or γ-irradiation. Upon UV-excitation the number of chains initiated is of order $10^{-2}$ per absorbed photon. The active precursor state is likely to be of triplet character. Even in case of optical generation librational excitation of the reacting molecules is required to cross over the potential barrier. Chain growth proceeds via consecutive first order 1.4-addition reactions each coupling one diacetylene monomer to a reactive intermediate. In the early stages of the reaction this intermediate is a diradical oligomer. It is converted to a dicarbene or an asymmetric carbene as the oligomer length increases beyond 5 units. The addition reaction is a thermally activated non-photochemical process. Its activation energy, typically 0.3 eV, is a linear function of the C1–C4 separation. If the length of the repeat unit of oligomer/polymer is shorter than the monomer lattice constant along $\vec{b}$, the activation energy increases with growing chain length. In this case the reaction rate can be enhanced by polymer chains already present and acting as springs to compress the structure of the reaction matrix thereby giving rise to autocatalytic reaction enhancement. On the basis of existing experimental data a diagram relating activation energy and reaction distance can be constructed for the reaction of TS-6. In essence it is the superposition of a harmonic vibrational potential and an electronic coupling energy and can be used to calculate the kinetic chain length. The result is in remarkably good agreement with the experimental values obtained in the low conversion limit ($X < 0.1$, $n \approx 18$ to 35 repeat units). In cases where lattice mismatch is unimportant, i.e., within the autocatalytic reaction regime or in materials where monomer and polymer are almost in register, the maximum kinetic chain length $n_\infty$ is either limited by the length of a row of monomers confined between the dead ends of previously formed chains or the number of addition events occurring per lifetime of the reactive chain end. For 4-BCMU and TS-12, $n_\infty$ is of order 2000. Several estimates indicate that in TS-6 the limiting chain length does not exceed a few hundred repeat units.


# 7 Secondary Chain Initiation in TS-6

Reaction-independent chain initiation as well as reaction-driven increase of the chain length by not more than one order of magnitude are incompatible with the experimental observation of a 200-fold increase of the total reaction rate (see section 2) in the autocatalytic reaction regime in TS. The discrepancy appears to be even more dramatic if photopolymerization is carried out under hydrostatic pressure. Upon UV-irradiation of the (1 0 0) area of TS-6 and TS-12 crystals through a silver coated mica mesh containing radiochemically fabricated rhombic holes of 10 μm edge length Braunschweig and Bässler [111] observed polymer growth into unirradiated crystal zones. It was found that the size of the polymerized area increases upon applying hydrostatic pressure. Interestingly, the effect turned out to be almost isotropic within the (1 0 0) plane (see Fig. 18). At $p \approx 4$ kbar the whole crystal surface became polymerized. Taking into account the area density of irradiation spots ($\sim 7 \cdot 10^3$ cm$^{-2}$) the individual polymerization zones must ultimately reach a diameter of about 60 μm. This is equivalent to chain growth across about 25 μm into unirradiated crystal areas, in apparent disagreement with both the single chain concept and the commonly accepted view that chains grow only along [0 1 0].

To resolve the paradox, it has been postulated [111] that a growing chain is able to initiate secondary chains in its neighborhood by virtue of the energy of 1.6 eV

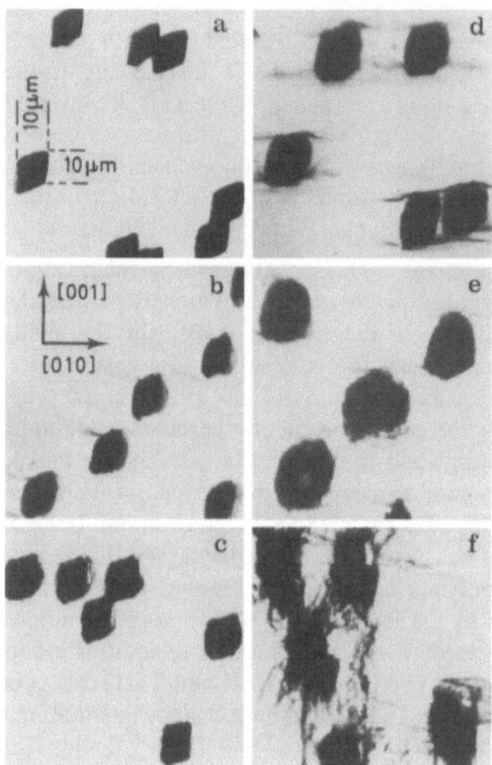

**Fig. 18a–f.** Microscopic views of the (1 0 0) plane of TS-crystals after UV-irradiation through a mica mesh containing rhombic holes and subsequent etching. Photographs were taken with light polarized parallel to the polymer chain axis. Parameter is the pressure during photopolymerization; **b**: 1 bar, **c**: 1 kbar, **d**: 1.5 kbar, **e**: 2 kbar, **f**: 2.5 kbar. **a** is a view of the mesh

released by the monomer addition reaction. For steric reasons it appears conceivable that the librational motion involved in this reaction couples to the motion of unreacted monomers in an adjacent stack of molecules. This concept can explain (i) why more monomer molecules are consumed per primary chain initiation event than comprised within an individual chain and (ii) why the polymerization zone also grows in a direction perpendicular to the chain. Because of the strong pressure dependence of the rate of the individual reaction step, the effect of secondary chain initiation must also be subject to enhancement by pressure in accord with the experimental result.

In principle, the concept of secondary chain growth is in conflict with the commonly accepted notion that polymer growth inside the reacting matrix is a random process not involving any nucleation effects. However, a recent estimate indicates that in practice this contradiction does not exist. Analyzing thermal conversion data, Braun-schweig and Bässler [112] concluded that the probability for initiating a new chain per monomer addition reaction is of order $10^{-2}$. Therefore, the probability that in the low conversion regime of TS-6, where n < 30, a secondary chain is created next to promotor chain is only 0.3. The situation changes not until the high conversion regime is reached. Here, the concept of independent chain growth becomes irrelevant, since a new chain will practically always be generated in the vicinity of an existing chain.

# 8 Photopolymerization of Multilayer Systems

## 8.1 Undoped Systems

Polymerization of diacetylenes is a lattice controlled topochemical reaction. Although guaranteeing well-ordered reaction products this fact imposes serious restrictions regarding chemical tailoring for optimizing product properties. It would, for instance, be highly desirable to prepare samples that combine the unique charge transport properties of crystalline polydiacetylenes with the large conductivity of doped polymers like polyacetylene or other conjugated systems [113-116]. Since the bandgap of a neat PDA chain is close to 2.5 eV [117-119], this would either require doping by electron donors or acceptors, or, preferably, introduction of substituents that undergo efficient charge transfer with the polymer chain either in the dark or after optical excitation. Except DCH which shows weak charge transfer from the carbazole group to the chain after optical excitation [120] none of such systems has been prepared so far, mainly because the structural modifications introduced in the monomer lattice turn out to be prohibitive for future reaction.

A major achievement towards greater flexibility with respect to structural constraints of the reaction is the discovery of polymerization in Langmuir-Blodgett (LB)-films made by Wegner and coworkers [121, 122]. These systems also allow introducing dye molecules acting as sensitizers for the polymerization reaction.

The LB-technique of growing mono- or multi-molecular layers is widely known [123] and shall not be repeated here. In general, the ability to form monolayer assemblies requires that the active molecules carry a hydrophilic head group, typically $-(CH_2)_nCOOH$, and a long hydrophobic tail, normally an n-paraffine chain, $-(CH_2)_m$. A list of substituted diacetylenes synthetized by Tieke et al. [124] with the aim of producing LB-films is presented in Table 2. It contains information regarding formation of mono- and multilayers and their ability to polymerize upon exposure to UV-irradiation ($\lambda < 300$ nm). Polymerization is detected by appearance of the polymer spectrum. Figure 19 shows absorption spectra recorded after irradiation of compound 3 for various exposure times. As polymerized films appear blue, i.e., they absorb in the red like most crystalline diacetylenes do. After treatment with ethanol or chloroform or by sample annealing at 90 °C, the absorption spectrum is blue-shifted irreversibly and the layer appears red. The color change is believed to result from a phase transition involving reorganization within the side groups in course of which the grain size of the layer structure is reduced by one order of magnitude [125, 126]. This effect shall not be considered further.

Comparison of the monomer-absorption spectrum and the photoaction spectrum for polymerization (Fig. 20) demonstrates that it is excitation of the diacetylene moiety which drives the photoreaction, in analogy to what is known from single crystal work. Upon varying the number of layers in a multilayer assembly a strictly linear relationship between optical density obtained in the saturation limit and layer thickness is observed. This documents that the multilayer structure is preserved in course of the polymerization process. Assuming quantitative conversion an optical density OD $= 4.56 \cdot 10^{-3}$ per monolayer corresponding to an absorption coefficient $\alpha = 1.61 \cdot 10^4$ cm$^{-1}$ is calculated for the red form absorbing at 500 nm. Existence of the layer structure is further verified by small-angle X-ray scattering. Although

**Table 2.** Spreading and polymerization behavior of long-chain-diacetylenes with ability to form monomolecular films

| Compound R—C≡C—C≡C—R' | | Ability to form monomolecular films | | Multilayer formation | Photopolymerization[a] ($\lambda < 300$ nm) | |
|---|---|---|---|---|---|---|
| R | R' | On $H_2O$ | On $10^{-3}$ M $CdCl_2$-sol | | In crystal | In multilayer |
| 1   n—$C_9H_{19}$— | —$(CH_2)_8$—COOH | Yes | Yes | | ++ | — |
| 2   n—$C_{10}H_{21}$— | —$(CH_2)_8$—COOH | Yes | Yes | Yes | ++ | +++ |
| 3   n—$C_{12}H_{25}$— | —$(CH_2)_8$—COOH | Yes | Yes | Yes | ++ | ++ |
| 4   n—$C_{14}H_{29}$— | —$(CH_2)_8$—COOH | Yes | Yes | Yes | ++ | + |
| 5   n—$C_{12}H_{25}$— | —$(CH_2)_3$—COOH | Yes | Yes | No | ++, +[b] | — |
| 6   n—$C_{16}H_{33}$— | —$(CH_2)_2$—COOH | Yes | Yes | Yes | ++ | — |
| 7   n—$C_{16}$—$H_{33}$— | —COOH | Yes | Yes | Yes | +++ | + |
| 8   n—$C_{12}H_{25}$— | —COOH | No | Yes | No | +++ | — |
| 9   n—$C_{10}H_{21}$— | —$(CH_2)_9$—OH | Yes | Yes | No | + | — |

[a] +++: Dark blue in less than 10 sec; ++: dark blue in 10–30 sec; +: dark red after exposure $>2$ min; —: no reaction;
[b] 5 has two modifications with different reactivity

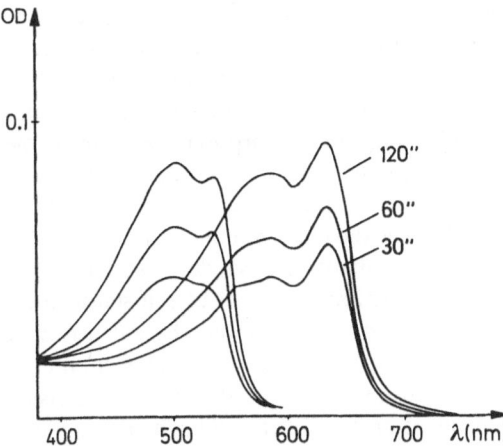

**Fig. 19.** Absorption spectra of a multilayer assembly containing 28 layers of the Cd-salt of n–$C_{12}H_{25}$–C≡C–C≡C–$(CH_2)_8$–COOH after various exposure times to UV-irradiation. The long-wave-length spectra refer to as polymerized films, blueshifted absorption spectra are obtained after treatment with ethanol (from Ref. [124])

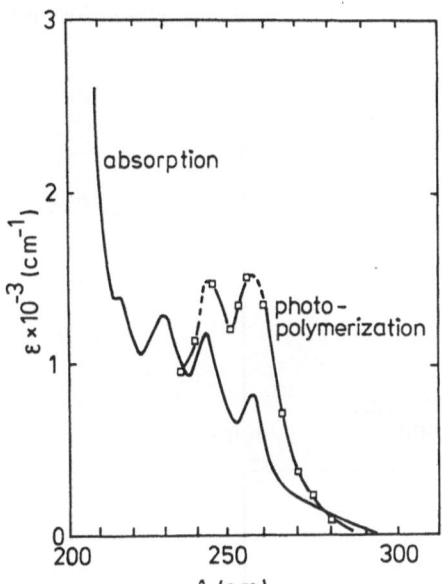

**Fig. 20.** Comparison between UV-absorption spectra of a paraffinsubstitited diacetylene (m = 8, n = 8) incorporated in a multilayer assembly and the action spectrum of photopolymerization (from Ref. [124])

free of holes, the sample is assumed to consist of domains with random orientation within the substrate plane. The domain size is estimated to be within the order of microns.

At low conversion the polymerization yield nq ranges between 5 and 10 [127]. Anticipating the primary quantum efficiency q to be similar to that in crystalline TS-6, i.e., q ∼ $10^{-2}$, a chain length of order 1000 repeat units would result. No autocatalytic reaction enhancement is observed. It appears that both the high degree of flexibility of the substituents and the way the molecule pack within the multilayer assembly provide sufficient motional freedom to accomodate strain developing in course of the reaction.

In course of subsequent work Bubeck, Tieke, and Wegner [128] discovered that the
action spectrum for photopolymerization of undoped diacetylene multilayers extends
into the visible provided some polymer formed in course of previous UV-irradiation
is present. Since obviously excitation of the polymer can sensitize the reaction this
effect has been termed self-sensitization. Checking the absorption spectrum of the
polymer produced via self-sensitization assured that the final product is identical
with the product obtained under UV excitation of the monomer. Later work by
Braunschweig and Bässler [111] demonstrated, that the effect is not confined to multilayer systems but is also present in partially polymerized single crystalline TS-6,
albeit with lower efficiency. Interestingly, the action spectrum of self-sensitization
follows the action spectrum for excitation of an electron from the valence band
of the polymer backbone to the conduction band [117] rather than the excitonic absorption spectrum of the polymer which is the dominant spectral feature in the visible
(see Fig. 21). The quantum yield is independent of the electric field, whereas in a one-
dimensional system the yield of free carriers, determined by thermal dissociation of
optically produced, weakly bound geminate electron-hole pairs, is an linear function
of an applied electric field [29, 30, 32, 129]. Apparently, the sensitizing action does not

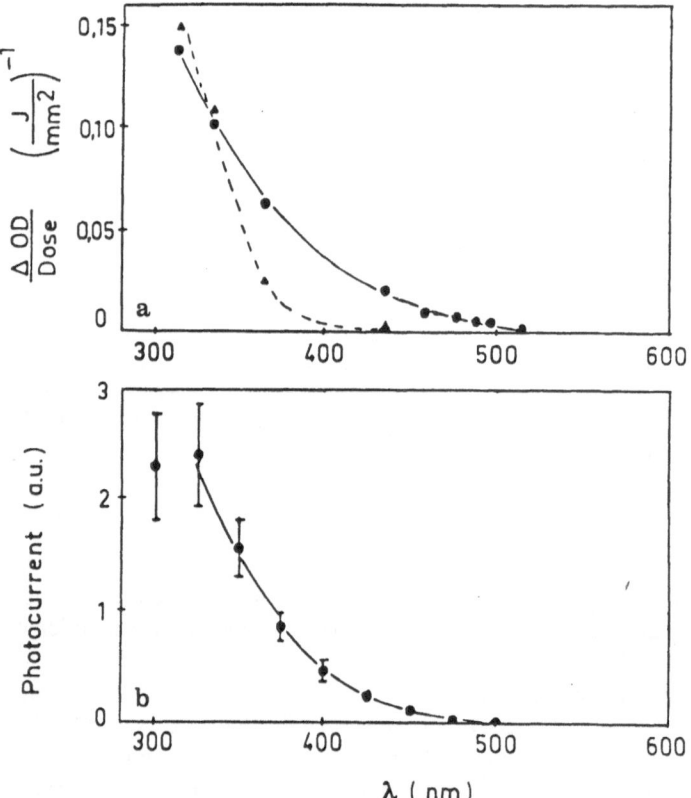

**Fig. 21. a.** Action spectra for photopolymerization of neat (triangles) and partially polymerized
(circles) multilayer systems. ΔOD is the change in optical density at a given irradiation dose; **b** Action
spectrum of intrinsic photoconductivity of a fully polymerized multilayer assembly [117] from Ref. [128]

require free carriers but can be brought about by electron-hole pairs which are likely to have a high mobility along the conjugated chain [130, 131].

The diacetylene monomer provides no energy states below 3.1 eV that could be populated via a sensitizer whose excitation is only 2.5 eV. Since its highest filled and lowest empty molecular level lie below and above the corresponding polymer levels, respectively, charge transfer from polymer to monomer is excluded as well. If, on the other hand, sensitization were due to secondary chain initiation involving local heating in course of a non-radiation transition, excitation of the polymer exciton state — known to decay non-radiatively — should be similarly effective, contrary to what is observed. The obvious conclusion is that self-sensitization reactivates dead ends of preformed polymer chains for further growth rather than generates new chains (see Fig. 22). Participation of charge carriers indicates that the essential process

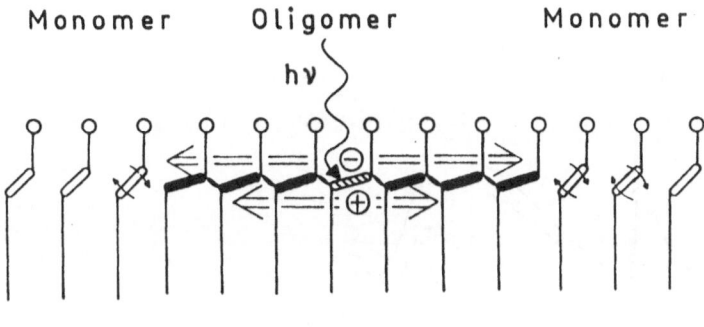

**Fig. 22.** Self-sensitization mechanism proposed by Bubeck et al. [128]

is an electronic rearrangement at the polymer head, whose structure has been suggested to be of the cyclopropene type [15]. It is conceivable that temporary lowering of the electron density by putting a positive charge of a coulombically electron-hole pair to the chain end breaks up the C1–C3 carbon bond and restores the carbene structure according to the scheme:

## 8.2 Doped Systems

Because of the dramatic change in color as well as mechanical stability upon polymerization diacetylenes appear to be attractive materials for image technology and
information storage. A serious constraint regarding device construction is the high
photon energy required to initiate photopolymerization. To overcome this difficulty
there was an intense effort for finding appropriate sensitizers that could shift the
photoresponse spectrum into the visible. The first material of this sort was a mixed
crystal composed of amphiphilic diacetylene monocarbonic acids and phenazine.
Tieke and Wegner [132] found that the 2:1 complex of 2,4-heptadecadiynoic acid and
phenazine (HD-Ph) polymerizes upon irradiating into the phenazine first singlet
absorption band near 400 nm (Fig. 23). Later on Tieke and Bloor [133] observed that

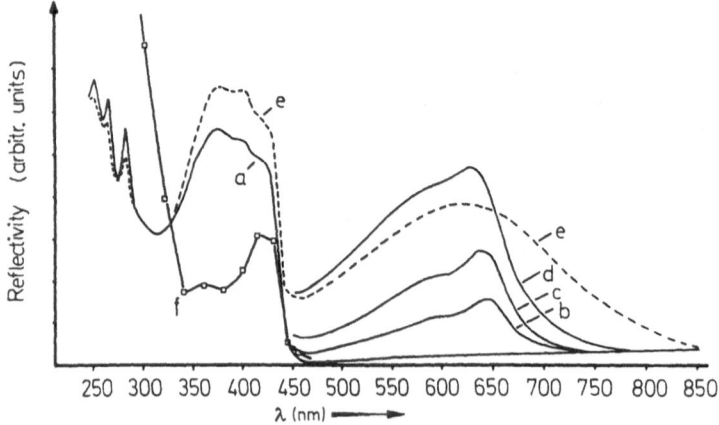

**Fig. 23.** Reflexion spectra of the 2:1 complex of 2.4 heptadecadiynoic acid and phenazine before (**a**),
and after sample irradiation at 410 nm for 5 min (**b**), 10 min (**c**) and 20 min (**d**). Spectrum e was
obtained after sample annealing at 90 °C. f is the action spectrum for photopolymerization at
$\lambda > 300$ nm in arbitrary units (from Ref. [132])

nonadeca — as well as heneicosa 2,4-diynoic acid — phenazine complexes behave
similarly. Monitoring resonant Raman spectra and X-ray diffraction pattern during
the reaction allowed elucidating the molecular reorganization processes accompanying
the reaction. Braunschweig and Bässler [65] investigated the reaction kinetics following
build-up of the polymer absorption as a function of irradiation time and concluded
on the active role played by the phenazine singlet state. The quantum yield (nq)
turned out to be 0.5 for direct excitation of the diacetylene moiety (280 nm) and 0.1
for phenazine excitation (420 nm). On the basis of ESR-studies Bubeck et al. [155]
have meanwhile established that the reaction is triggered by charge transfer from
diacetylene to the excited phenazine.
    The range of potential sensitizers for photopolymerization of diacetylenes widened
considerably when Wegner and coworkers [134, 135] discovered sensitized photopolymerization in LB-multilayer systems doped with certain dyes. It is known for long
that dye molecules carrying long aliphatic substituents can easily be incorporated into
LB-multilayer assemblies without destroying the sample architecture [136]. Therefore,
the topotactic requirements for the polymerization reaction to proceed are retained.

**Table 3.** List of dyes used as sensitizers for photopolymerization of diacetylene multilayer assemblies. (R: $-(CH_2)^1{}_7-CH_3$; the counterion was $I^-$) and their absorption maxima

| | Structure | $\lambda_{max}(nm)$ |
|---|---|---|
| I | 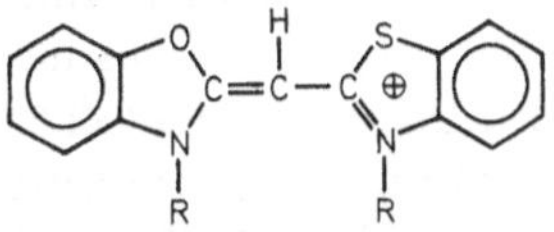 | 408 |
| II | 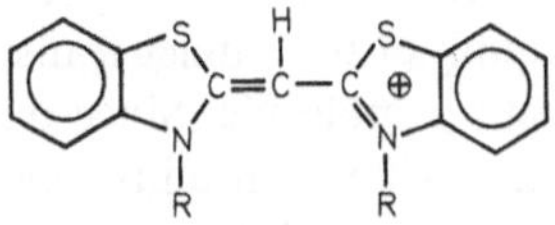 | 430 |
| III | 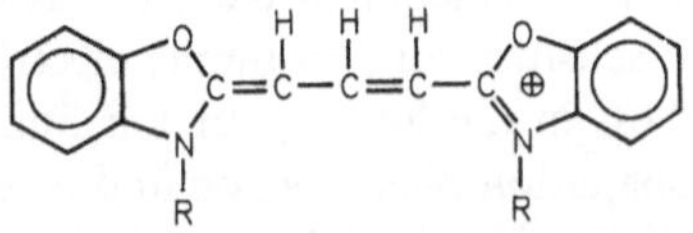 | 498 |
| IV | 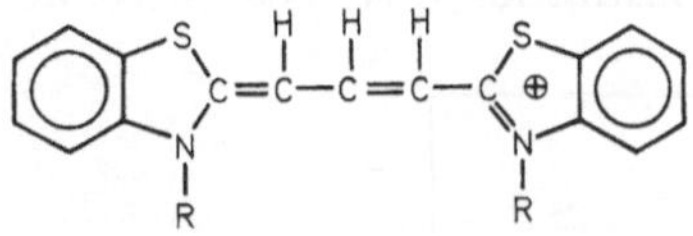 | 572 |
| V | 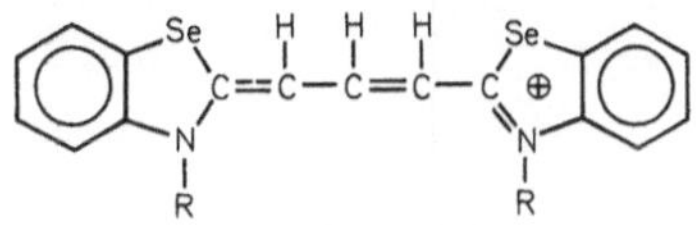 | 586 |

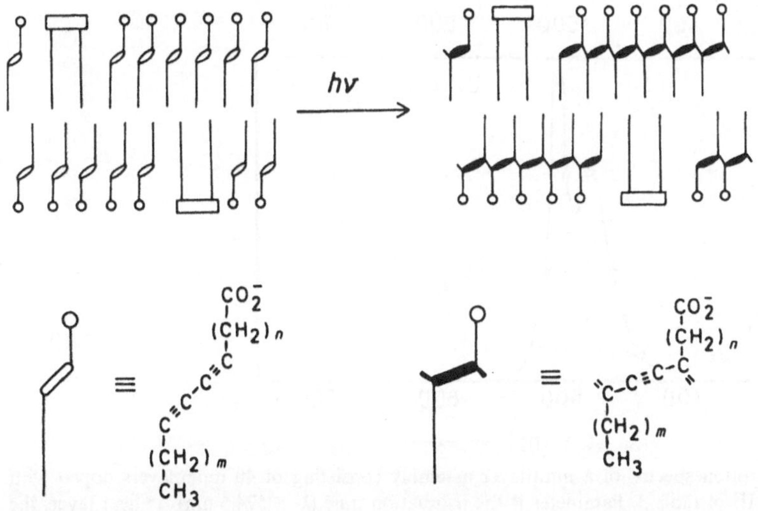

**Fig. 24.** Structural model of monomeric and polymeric multilayer assemblies doped with cyanine dyes. The black bars represent the polydiacetylene $\pi$-bond system (from Ref. [135])

Table 3 presents a list of molecules used by Bubeck at al. [135] to sensitize polymerization of amphiphilic diacetylene multilayers. A model of the layer architecture is shown in Fig. 24. Upon excitation into the absorption band of the chromophore the absorption spectrum characteristic of the polymer gradually develops and concomitantly the dye fluorescence becomes quenched (Fig. 25). The final product is identical with the product obtained upon UV-excitation of the diacetylene. Although sensitization occurs with a low quantum efficiency — 1400 ± 900 photons have to be absorbed by a dye molecule to convert one monomeric diacetylene molecule into a polymer unit — its yield Y expressed in total number of monomers converted to polymer per dye molecule is large. Since a sample containing only $2 \cdot 10^{-4}$ mole/mole dye molecules is fully polymerized after prolonged pumping of the singlet transition of the sensitizer, Y must be >5000 monomers per sensitizer molecule. Since on the other hand, the total chain length is undoubtedly less than 5000 repeat units, one dye molecule must be capable of sensitizing repeatedly. This is incompatible with any irreversible changes of the dye and demonstrates that sensitization is a purely electronic phenomenon. Since its yield turns out to be highest in the low conversion limit, it must be a chain initiation rather than a chain propagation phenomenon and as such be principally different from the effect of self-sensitization described in section 8.1.

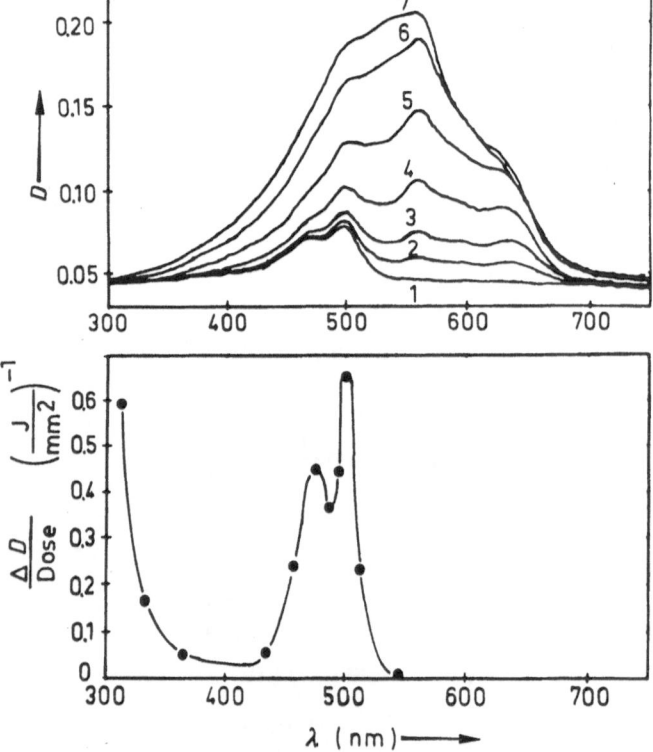

**Fig. 25. a**: Absorption spectra of a multilayer assembly consisting of 40 monolayers doped with 1 mol.-% of dye III of table 3. Parameter is the irradiation time ($\lambda = 574.5$ nm). 1: neat layer, the absorption spectrum is that of the dye, 2: 2 s, 3: 20 s, 4: 1.5 min, 5: 4.5 min, 6: 12 min, 7: 24 min; **b** Action spectrum of sensitized photopolymerization (from Ref. [135])

The low excitation energy of the sensitizer as compared to the diacetylene monomer rules out energy transfer to be the rate determining step for chain initiation. Instead, either charge transfer from sensitizer to diacetylene or transfer of vibronic energy released at the sensitizer as a result of non-radiative deactivation have to be invoked. The second alternative can be discarded since Bubeck et al. [137] found no systematic difference on the sensitizing efficiency of fluorescent and nonfluorescent dyes. Therefore, one has to postulate a charge transfer mechanism as the rate determining step for promoting sensitized chain initiation.

Sensitization by charge transfer requires that the LUMO of the diacetylene monomer be below the excited singlet level of the dye. The latter is at $I_s - E_{S1}$ where $I_s$ and $E_{S1}$ are ionization potential and energy of the first singlet state of the dye, respectively. On the basis of polarographic data $I_s - E_{S1} \sim -2.7 \ldots -3.3$ eV is estimated for cyanine dyes [138]. The lowest unoccupied orbital of a previously neutral diacetylene molecule is at $A + P^-$ where A is the electron affinity and $P^-$ the polarization energy of a molecular anion. In molecular crystals a typical value for $A + P^-$ is $-2.0$ eV [139]. Since in course of a CT-process a coulombically bound electron-hole pair is generated at a pair of adjacent molecules a coulombic binding energy $E_{coul}$ of order $-1$ eV has to be considered in the energy balance. In view of the uncertainty of these data one can safely conclude that the energetic condition for charge transfer from sensitizer to diacetylene to occur, $A + P^- + E_{coul} < I_s - E_{S1}$, is likely to be fulfilled for the dye molecules under consideration.

In summary, the results on sensitized photopolymerization of diacetylenes are of relevance for potential application and, in addition, augment understanding of the polymerization process in general. In conjunction with what is known about chain propagation and self-sensitization, they indicate that the basic requirements for both chain initiation and propagation are presence of an unpaired electron — in form of either a radical or a charge carrier — on the reactive center and excitation of a librational motion of the reactant(s) in course of which the reaction distance is temporarily reduced to a critical value.


# 9 Application

Polymerization of crystalline diacetylenes generates an ordered array of fully extended polymer chains with conjugated backbone. The increase of the extent of the $\pi$-electron system, initially confined to the diacetylene linkage, over many polymer repeat units causes a dramatic lowering of the excitation energy manifested in a shift of the optical absorption into the red. For this reason diacetylenes appear to be attractive materials to be used as active elements in devices for optical information storage or photolithography. The basic advantages as well as disadvantages faced upon such application are easily gleaned from the available data for photochemical yield and chain length. Since the primary quantum efficiency for chain initiation is of order $10^{-2}$, a total-gain $>1$ can only be achieved if the total number of reached monomers per photon is $>100$. This requires materials in which reaction is not impeded by mismatch between monomer and polymer lattices, such as 4-BCMU, where $n \approx 2400$ and $nq \sim 60$ at room temperature [74]. Unfortunately, the reaction requires UV light of $\lambda \lesssim 330$ nm to get started. Use of sensitizers, for instance, dye molecules incorporated into multilayer assemblies, which extend the spectral range into the visible, do so at a drastic loss of efficiency. Therefore, diacetylenes do not appear to be suitable materials for imaging technology that requires large efficiencies. Nevertheless, their use as photographic receptors has been reported [140]. On the other hand, the moderate length of the polymer product allows image recording at a spatial resolution far beyond the resolution of conventional receptor systems, relevant, e.g., in microlithography. In a neat TS crystal, for example, which, if carefully grown, containing less than 0.1 percent polymer, the length of a developing polymer chain is limited to about 100 Å. In this case it would be the spatial resolution of the writing beam rather than that of the receptor material that controls the resolution of the image. It seems attractive to speculate about the use of ultra narrow electron beams for writing-in image patterns which could subsequently be stabilized against undesired secondary polymerization by etching. Preliminary experiments in this laboratory have indicated that in this case it is the spread caused by electron diffusion within the monomer crystal which limits spatial resolution.

An interesting application of the high spatial resolution power associated with photopolymerization of diacetylenes is holographic image recording recently studied by Richter, Güttler, and Schwoerer [141, 142]. These authors irradiated a high quality TS-6 single crystal with two interfering plane wave beams obtained from a frequency doubled Ar-laser ($\lambda = 257$ nm) by beam splitting (see Fig. 26). In the region of con-

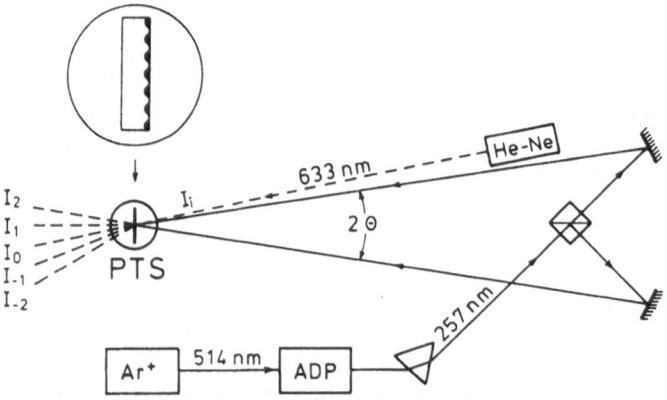

**Fig. 26.** Experimental arrangement for recording and reactions of a plane wave hologram on the surface of a TS-crystal (from Ref. [141])

structive interference the crystal surface polymerizes up to a depth of about 1 μm which is the penetration depth of 257 nm light in the monomer crystal. The interference pattern is a grating with equidistant lines, $\Lambda = \lambda/(2 \sin \theta)$ apart, where $2\theta$ is the angle between the interfering beams. Variation of $2\theta$ between 3° and 16° gave line distances between 5 and 0.9 μm. An example of a grating produced after 2 s recording time at a incident power of 60 mW and viewed under the transmission microscope in white light is shown in Fig. 27. Since under low conversion conditions

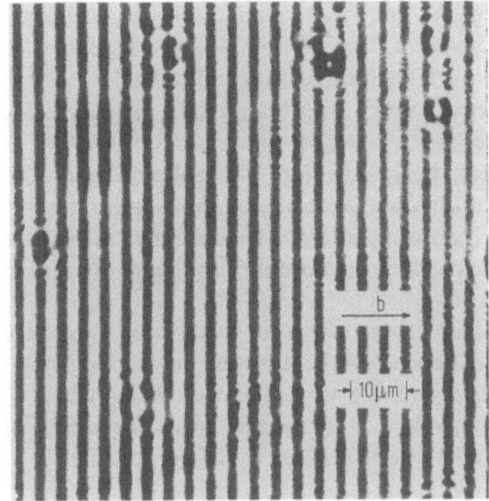

**Fig. 27.** Holographic grating on the surface of a TS crystal. Dark areas indicate polymerized TS (from Ref. [141])

the chain length in TS is of order 100 Å, the width of the polymerized lines is determined by the spatial extent of the regions of constructive interference between the recording beams. Depending upon the wavelength of the read-out light the recorded image acts as both a phase and amplitude grating. If read-out is done with a HeNe-laser operating at 633 nm where polymer absorption is weak the diffraction effect arising from the phase hologram prevails. Diffraction efficiencies, defined as the ratio of the first order diffraction intensity to the incident intensity of the analysing light up to 35 percent were observed at a exposure light dose of about 100 mJ/cm$^2$.

Holographic gratings produced in the above manner need not be processed after recording. However, a weak hologram can be developed by gentle annealing in the dark to increase its efficiency. In this case the autocatalytic enhancement of thermal polymerization in TS once the conversion has reached about 10 percent is utilized as a stimulus for further growth of the polymeric grating lines [142].

Another potential application of fully converted polydiacetylenes is based upon their unusual non-linear optical properties. Sauteret et al. [143] found that the third order susceptibilities of TCDU and TS increases by about a factor of 600 upon polymerization and become comparable to those of inorganic semiconductors like GaAs or germanium. This is a consequence of the increase of $\pi$-electron delocalization upon polymerization. Polymeric diacetylenes can therefore be used as efficient elements for third harmonic generation. In Ref. [143] this effect has been employed for tripling the frequencies of 1.89 µm and 2.62 µm radiation.

The unusual non-linear optical properties of polymer diacetylenes opened attractive prospects for their application in three or four-wave parametric amplifiers. Sauteret et al. [143] reported that with a 1.06 µm, 1 GW/cm$^2$ pump the gain for four-wave parametric amplification in the near infrared in PTS is about ten times larger than the gain for three-wave parametric amplification in LiNbO$_3$. However, later work [144,

[145)] showed that usage of polydiacetylenes in three-wave mixing[2] devices is affected by strong two-quantum absorption. Therefore, the spectral window available for application is limited by onset of intense vibrational transitions at the high wavelength side ($\sim 3 \mu m$) and onset of two- or three-quantum absorption of the primary beam at the short wavelength side.

The color change of diacetylenes upon polymerization, either thermally or UV and $\gamma$-induced, can be used in temperature-dose or radiation-dose indicators [146]. Their advantage is their high sensitivity allowing to cover a dose range of 50 rads to 50 Mrads. In addition, some compounds undergo an abrupt color change in solution at a certain concentration [85, 147, 148].

If different diacetylenes are cocrystallized they may form reactive or unreactive phases. Which of these actually occur, can depend on the substrate on which crystallization proceeds. For instance. spraying a mixture of a phenylurethane and a p-chlorophenylurethane diacetylene into an oil-covered surface causes formation of the unreactive modification. This effect has been utilized by Miller and Patel [149] to record latent fingerprints.

As far as their electrical properties are concerned, polydiacetylenes are mainly used as model systems for basic research on one-dimensional crystalline solids that are non-conductive under conditions of thermodynamic equilibrium, although potential technical applications have also been suggested. For instance, there is considerable commercial interest for fabricating thin well-defined insulating layers to be used in metal-insulator-semiconductor devices. Owing to their high resistivity [150] and well defined thickness, monolayer-assemblies are potential candidates. Unfortunately, the molecular architecture of conventional fatty acid systems renders them highly vulnerable to mechanical damage. Considerable improvement of the stability of such films has been made by the discovery of polymerizable diacetylene LB-layer systems. They exhibit the good insulating properties characteristic of fatty acid systems and deposit readily on semiconductor surfaces. These promising features prompted Roberts and coworkers [151–153] to investigate the electrical properties of polymeric LB layers in detail. They found, that polymeric diacetylene LB layers can be successfully deposited, e.g., onto the narrow band gap semiconductors InSb and (Hg-Cd)Te that do not possess high quality native oxides like $SiO_2$ and are of interest for fabricating IR-detectors. The electrical performance of MIS structures showed that the LB films may be used for either passivating the surface of photoconduction devices or provide the basis for an MIS technology for narrow band gap semiconductors. Although the results appear encouraging the domain boundaries present in the LB assemblies constitute a potential source of leakage currents that might impose a restriction regarding stability and reproducibility of devices produced on a commercial scale.

Although being insulators in general at least one crystalline polydiacetylene DCH was found to be electrochemically dopable up to a conductivity of $10^{-3}$ $(\Omega m)^{-1}$ by applying an $NaJ/J_2$ electrolytic contact [154]. One could think of combining this technique with electron-beam polymerization in order to produce a well-defined conducting pattern at the surface of a diacetylene crystal support or LB system.

---

2 In a three-wave mixing experiment, two incident beams of frequency $w_1$ and $w_2$ are focussed onto the mixing element and an outcoming beam generated at $w_3 = 2w_1 - w_2$ is detected.

# 10 Concluding Remarks

It appears that the basic mechanisms involved in polymerization and, in particular, in photopolymerization of diacetylenes are understood. Nevertheless, to date it is not possible to design a diacetylene monomer on the basis of desired reaction behavior and/or product properties. To this end more quantitative information on the reaction kinetics of diacetylenes including quantum-mechanical aspects is needed which could form the basis for developing a quantitative relationship between structure and reactivity.

*Acknowledgement*: I am indebted to F. Braunschweig for numerous discussions and to D. Bloor, R. R. Chance, G. G. Roberts, M. Schwoerer, H. Sixl, and G. Wegner for communicating their results prior to publication and/or helpful comments. Financial support by the Stiftung Volkswagenwerk and the Fonds der Chemischen Industrie for part of this work is gratefully acknowledged.

# 11 References

1. Wegner, G.: Z. Naturforsch. *24B*, 824 (1969)
2. Wegner, G.: Makromol. Chem. *145*, 85 (1971)
3. Kaiser, J., Wegner, G., and Fischer, E. W.: Israel. J. Chem. *10*, 157 (1972)
4. Wegner, G.: Makromol. Chem. *154*, 35 (1972)
5. Wegner, G. in: "Molecular Metals", W. E. Hatfield (ed.), Plenum Press, N.Y. (1979), p. 209
6. Hirshfeld, F. L. and Schmidt, G. J. M.: J. Polym. Sci. Part A *2*, 2181 (1964)
7. Schmidt, G. M. J.: "Photochemistry of the Solid State", in "Reactivity of the Photoexcited Organic Molecule", Wiley, N.Y. (1967), p. 227
8. Baughman, R. H.: J. Appl. Phys. *43*, 4362 (1972)
9. Baughman, R. H.: J. Polym. Sci. Polym. Phys. Ed. *12*, 1511 (1974)
10. Baughman, R. H. and Yee, K. C.: J. Polym. Sci. Polym. Chem. Ed. *12*, 2467 (1974)
11. Baughman, R. H. and Chance, R. R.: Ann. Acad. Sci. N.Y. *313*, 705 (1978)
12. Schermann, W., Wegner, G., Williams, J. O., and Thomas, J. M.: J. Polym. Sci. Polym. Phys. Ed. *13*, 753 (1975)
13. Sixl, H., Hersel, W., and Wolf, H. C.: Chem. Phys. Letts. *53*, 39 (1978)
14. Bubeck, C., Neumann, W., and Sixl, H.: Chem. Phys. *48*, 269 (1980)
15. Hersel, W., Sixl, H., and Wegner, G.: Chem. Phys. Letts. *73*, 280 (1980)
16. Neumann, W. and Sixl, H.: Chem. Phys. *58*, 303 (1981)
17. Gross, H., Neumann, W., and Sixl, H.: Chem. Phys. Letts. *95*, 584 (1983)
18. Stevens, G. C. and Bloor, D.: Chem. Phys. Letts. *40*, 37 (1976)
19. Eichele, H., Schwoerer, M., Huber, R., and Bloor, D.: Chem. Phys. Letts. *42*, 342 (1976)
20. Huber, R., Schwoerer, M., Bubeck, C., and Sixl, H.: Chem. Phys. Letts. *53*, 35 (1978)
21. Bubeck, C., Sixl, H., and Wolf, H. C.: Chem. Phys. *32*, 231 (1978)
22. Hori, Y. and Kispert, D.: J. Am. Chem. Soc. *101*, 3173 (1979)
23. Huber, R. and Schwoerer, M.: Chem. Phys. Letts. *72*, 10 (1980)
24. Schwoerer, M., Huber, R. A., and Hartl, W.: Chem. Phys. *55*, 97 (1981)
25. Huber, R. A., Schwoerer, M., Benk, H., and Sixl, H.: Chem. Phys. Letts. *78*, 416 (1981)
26. Baughman, R. H. and Chance, R. R.: J. Polym. Sci. Polym. Phys. Ed. *14*, 2037 (1976)
27. Exarhos, G. J., Risen, W. M., Jr., and Baughman, R. H.: J. Am. Chem. Soc. *98*, 481 (1976)
28. Lochner, K., Reimer, B., and Bässler, H.: Phys. Letts. *41*, 388 (1976)
29. Lochner, K., Reimer, B., and Bässler, H.: phys. stat. sol. (b) *76*, 533 (1976)
30. Siddiqui, A. S. and Wilson, E. G.: J. Phys. C *12*, 4237 (1979)
31. Donovan, K. J. and Wilson, E. G.: Phil. Mag. B *44*, 31 (1981)
32. Seiferheld, U., Ries, B., and Bässler, H.: J. Phys. C. *16*, 5189 (1983)
33. Hunt, I. G., Bloor, D., and Movaghar, B.: J. Phys. C., *16*, L 623 (1983)

34. Seiferheld, U., Bässler, H., and Movaghar, B.: Phys. Rev. Letts. *51*, 813 (1983)
35. Bloor, D., Ando, D. J., Preston, F. H., and Stevens, G. C.: Chem. Phys. Letts. *24*, 407 (1974)
36. Reimer, B., Bässler, H., Hesse, J., and Weiser, G.: phys. stat. sol. (b) *73*, 709 (1976)
37. Bloor, D. and Preston, F. H.: phys. stat. sol. (a) *37*, 427 (1976)
38. Bloor, D. and Preston, F. H.: phys. stat. sol. (a) *39*, 607 (1977)
39. Hood, R. J., Müller, H., Eckhardt, C. J., Chance, R. R., and Yee, K. C.: Chem. Phys. Letts. *54*, 295 (1978)
40. Sebastian, L. and Weiser, G.: Chem. Phys. *62*, 447 (1981)
41. Wegner, G., in: "Chemistry and Physics of the One-Dimensional Metals", Keller, H. J. (ed.), Plenum Press, N.Y. (1977), p. 297
42. Baughman, R. H. and Yee, K. C.: J. Polym. Sci. Macromol. Revs. *13*, 219 (1978)
43. Wegner, G.: Disc. Farad. Soc. *68*, 494 (1980)
44. Enkelmann, V., in: "Quantum Theory of Polymers", Springer Lecture Notes in Physics *113*, 1 (1980)
45. Bloor, D., in: "Quantum Theory of Polymers", Springer Lecture Notes in Physics *113*, 14 (1980)
46. Bloor, D.: NATO summerschool on Quantum Theory of Polymers, Braunlage, 1983, to be published
47. Batchelder, D. N. and Bloor, D.: Adv. in Infrared and Raman Spectroscopy, Clark, R. J. H. and Hester, R. E. (eds.), Vol. 11, 133 (1983)
48. Reimer, B. and Bässler, H.: phys. stat. sol. (a) *32*, 435 (1975)
49. Chance, R. R. and Baughman, R. H.: J. Chem. Phys. *64*, 3889 (1976)
50. Takura, Y., Mitani, T., and Koda, T.: Chem. Phys. Letts. *75*, 324 (1980)
51. Chance, R. R. and Sowa, J. M.: J. Am. Chem. Soc. *99*, 6703 (1977)
52. Baughman, R. H.: J. Chem. Phys. *68*, 3110 (1978)
53. Chance, R. R. and Patel, G. N.: J. Polym. Sci. Polym. Phys. Ed. *16*, 859 (1978)
54. Yee, K. C. and Chance, R. R.: J. Polym. Sci. Polym. Phys. Ed. *16*, 431 (1978)
55. Enkelmann, V., Leyrer, R. J., Schleier, G., and Wegner, G.: J. Mat. Sci. *15*, 168 (1980)
56. Enkelmann, V.: Makromol. Chem. *179*, 2811 (1978)
57. Enkelmann, V.: J. Mat. Sci. *15*, 951 (1980)
58. Galiotis, C., Young, R. J., Ando, D. J., and Bloor, D.: Makromol. Chem. *184*, 1083 (1983)
59. Tieke, B., Bloor, D., and Young, R. J.: J. Mat. Sci. *17*, 1156 (1982)
60. Bloor, D., Ando, J. D., Fischer, D. D., and Hubble, C. L., in: "Molecular Metals", Hatfield, W. E. (ed.), Plenum Press, N.Y. (1979), p. 249
61. Ando, D. J., Bloor, D., Hubble, C. L., and Williams, R. L.: Makromol. Chem. *181*, 453 (1980)
62. Patel, G. N., Khanna, Y. P., Ivory, D. M., Sowa, J. M., and Chance, R. R.: J. Polym. Sci. Polym. Phys. Ed. *17*, 899 (1979)
63. Tieke, B. and Wegner, G.: Makromol. Chem. *179*, 1639 (1978)
64. Wolf, H. C. and Port, H.: J. Lum. *12/13*, 33 (1976)
65. Braunschweig, F. and Bässler, H.: Ber. Bunsenges. Phys. Chem. *84*, 177 (1980)
66. Prock, A., Shand, M. L., and Chance, R. R.: Macromol. *15*, 238 (1982)
67. Niederwald, H. and Schwoerer, M.: Z. Naturforsch. *38a*, 749 (1983)
68. Takabe, T., Tanaka, M., and Tanaka, J.: Bull. Chem. Soc. Japan *47*, 192 (1974)
69. Galiotis, C., Young, R. J., and Batchelder, D. N.: J. Polym. Sci. Polym. Phys. Ed., in press
70. Bhattacharjee, H. R. and Patel, G. N.: J. Photochemistry *16*, 85 (1981)
71. Niederwald, H., Richter, K.-H., Gütler, W., and Schwoerer, M.: Laser Chem. (1983), in press
72. Braunschweig, F. and Bässler, H.: in preparation
73. Prock, A., Shand, M. L., and Chance, R. R.: J. Chem. Phys. *76*, 583 (1982)
74. Eckhardt, H., Prusik, T., and Chance, R. R.: Macromol. *16*, 732 (1983)
75. Bloor, D., Korski, L., Stevens, G. S., Preston, F. H., and Ando, D. J.: J. Mat. Sci. *10*, 1678 (1975)
76. Grimm, H., Axe, J. D., and Kröhnke, C.: Phys. Rev. B *25*, 1709 (1982)
77. Batchelder, D. N. and Bloor, D.: J. Phys. C *11*, L629 (1978)
78. Bloor, D., Williams, R. L., and Ando, D. J.: Chem. Phys. Letts. *78*, 67 (1981)
79. Albouy, P. A., Patillon, D. N., and Pouget, J. P.: Mol. Cryst. Liq. Cryst. *93*, 239 (1983)
80. Enkelmann, V., Leyrer, R. J., and Wegner, G.: Makromol. Chem. *180*, 1787 (1979)
81. Stevens, G. C. and Bloor, D.: J. Polym. Sci. Polym. Phys. Ed. *13*, 2411 (1975)

82. Mondong, R. and Bässler, H.: Chem. Phys. Letts. *78*, 371 (1981)
83. Avakian, P. and Merrifield, R. E.: Phys. Rev. Letts. *13*, 541 (1964)
84. Patel, G. N. and Walgh, E. K.: J. Polym. Sci. Polym. Letts. Ed. *17*, 203 (1979)
85. Patel, G. N., Chance, R. R., and Witt, J. D.: J. Chem. Phys. *70*, 4387 (1979)
86. Wenz, G. and Wegner, G.: Makromol. Chem. Rapid. Comm. *3*, 231 (1982)
87. Wenz, G. and Wegner, G.: Mol. Cryst. Liq. Cryst. *96*, 98 (1983)
88. Wenz, G.: thesis, Freiburg 1983
89. Kiess, H. and Clarke, R.: phys. stat. sol. (a) *49*, 133 (1978)
90. Xiao, D. Q., Ando, D. J., and Bloor, D.: Chem. Phys. Letts. *90*, 247 (1982) and Mol. Cryst. Liq. Cryst. *93*, 201 (1983)
91. Bertault, M., Schott, M., and Sworakowski, J.: preprint
92. Enkelmann, V. and Wegner, G.: Angew. Chem. *89*, 432 (1977)
93. Lochner, K. and Bässler, H.: Ber. Bunsenges. Phys. Chem. *84*, 880 (1980)
94. Bloor, D., Kennedy, R. J., and Batchelder, D. N.: J. Polym. Sci. Polym. Phys. Ed. *17*, 1355 (1979)
95. Bloor, D.: Mol. Cryst. Liq. Cryst. *93*, 183 (1983)
96. Lochner, K., Bässler, H., and Hinrichsen, T.: Ber. Bunsenges. Phys. Chem. *83*, 899 (1979)
97. Chance, R. R., Patel, G. N., Turi, E. A., and Khanna, Y. P.: J. Am. Chem. Soc. *100*, 1307 (1978)
98. Chance, R. R. and Shand, M. L.: J. Chem. Phys. *72*, 948 (1980)
99. Wegner, G.: Makromol. Chem. *134*, 219 (1970)
100. Leyrer, R. J. and Wegner, G.: Ber. Bunsenges. Phys. Chem. *83*, 470 (1979)
101. Niederwald, H., Eichele, H., and Schwoerer, M.: Chem. Phys. Letts. *72*, 242 (1980)
102. Kröhnke, C., Enkelmann, V., and Wegner, G.: Chem. Phys. Letts. *71*, 38 (1980)
103. Baughman, R. H. and Chance, R. R.: J. Chem. Phys. *73*, 4113 (1980)
104. Siebrand, W.: J. Chem. Phys. *47*, 2411 (1967)
105. Cade, N. A. and Movaghar, B.: J. Phys. C *16*, 539 (1983)
106. Fuchs, C. F., Heisel, F., Voltz, R., and Coche, A., in: "Organic Scintillators and Liquid Scintillation Counting", Horrocks, D. L. and Peng, C. (eds.), Academic Press, N.Y. (1971), p. 171
107. Bertault, M., Fave, J. L., and Schott, M.: Chem. Phys. Letts. *62*, 161 (1979)
108. Birks, J. B.: "Photophysics of Aromatic Molecules", Wiley, N.Y. (1968)
109. Chance, R. R. and Baughman, R. H.: Eighth Molecular Crystal Symp. Santa Barbara, California, 199, Abstracts p. 181
110. Enkelmann, V., Schleier, G., Wegner, G., Eichele, H., and Schwoerer, M.: Chem. Phys. Letts. *52*, 314 (1977)
111. Braunschweig, F. and Bässler, H.: Chem. Phys. Letts. *90*, 41 (1982)
112. Braunschweig, F. and Bässler, H.: Mol. Cryst. Liq. Cryst. *96*, 153 (1983)
113. Chiang, C. K., Finscher, C. R., Park, Y. W., Heeger, A. J., Shirakawa, H., Lewis, E. J., Gau, G. C., and MacDiarmid, A. G.: Phys. Rev. Letts. *39*, 1098 (1977)
114. Ivory, D. M., Miller, G. G., Sowa, J. M., Shacklette, L. W., Chance, R. R., and Baughman, R. H.: J. Chem. Phys. *71*, 1506 (1979)
115. Shacklette, L. W., Chance, R. R., Ivory, D. M., Miller, G. G., and Baughman, R. H.: Synthetic Metals *1*, 307 (1980)
116. Rabolt, J. F., Kanazawa, K. K., Reynolds, J. R., and Street, G. B.: J. Chem. Soc. Chem. Comm. 347 (1980)
117. Lochner, K., Bässler, H., Tieke, B., and Wegner, G.: phys. stat. sol. (b) *88*, 653 (1978)
118. Donovan, K. J. and Wilson, E. G.: J. Phys. C *12*, 4857 (1979)
119. Sebastian, L. and Weiser, G.: Phys. Rev. Letts. *46*, 1156 (1981)
120. Lochner, K., Bässler, H., Sebastian, L., Weiser, G., Wegner, G., and Enkelmann, V.: Chem. Phys. Letts. *78*, 366 (1981)
121. Tieke, B., Wegner, G., Naegele, D., and Ringsdorf, H.: Angew. Chem. Int. Ed. Engl. *15*, 764 (1976)
122. Tieke, B., Graf, H. J., Wegner, G., Naegele, B., Ringsdorf, H., Banerjie, S., Day, D., and Lando, J. B.: Colloid. Polym. Sci. *225*, 521 (1977)
123. Kuhn, H., Möbius, D., and Bücher, H., in: "Physical Methods of Chemistry", Weissberger, A. and Rossiter, B. (eds.), Wiley, N.Y. (1972), Vol. I, part. IIIb, chapter VII
124. Tieke, B., Lieser, G., and Wegner, G.: J. Polym. Sci. Polym. Chem. Ed. *17*, 1631 (1979)

125. Lieser, G., Tieke, B., and Wegner, G.: Thin solid films *68*, 77 (1980)
126. Tieke, B. and Bloor, D.: Makromol. Chem. *180*, 2275 (1979)
127. Tieke, B. and Wegner, G.: Makromol. Chem. *179*, 1639 (1978)
128. Bubeck, C., Tieke, B., and Wegner, G.: Ber. Bunsenges. Phys. Chem. *86*, 495 (1982)
129. Ries, B., Schönherr, G., Bässler, H., and Silver, M.: Phil. Mag. B *48*, 93 (1983)
130. Donovan, K. J. and Wilson, E. G.: Phil. Mag. B *44*, 9 (1981)
131. Spannring, W. and Bässler, H.: Chem. Phys. *84*, 54 (1981)
132. Tieke, B. and Wegner, G.: Makromol. Chem. *179*, 2573 (1978)
133. Tieke, B. and Bloor, D.: Makromol. Chem. *182*, 133 (1981)
134. Fouassier, J. P., Tieke, B., and Wegner, G.: Israel J. Chem. *18*, 227 (1979)
135. Bubeck, C., Tieke, B., and Wegner, G.: Ber. Bunsenges. Phys. Chem. *86*, 499 (1982)
136. Kuhn, H. and Möbius, D.: Angew. Chem. *83*, 672 (1971)
137. Bubeck, C., Tieke, B., and Wegner, G.: Mol. Cryst. Liq. Cryst. *96*, 109 (1983)
138. Large, R. F., in: "Photographic Sensitivity", R. J. Cox (ed.), Academic Press (1973), p. 241
139. Silinsh, E. A.: "Organic Molecular Crystals", Springer Series in Solid State Sciences, Vol. 16 (1980)
140. Sohn, J. E., Garito, A. F., Desai, K. N., Narang, R. S., and Kuzyk, M.: Makromol. Chem. *180*, 2975 (1979)
141. Richter, K. H., Güttler, W., and Schwoerer, M.: Chem. Phys. Letts. *92*, 4 (1982)
142. Richter, K. H., Güttler, W., and Schwoerer, M.: Appl. Phys. A *32*, 1 (1983)
143. Sauteret, C., Hermann, J.-P., Frey, R., Pradère, F., Ducuing, J., Baughman, R. H., and Chance, R. R.: Phys. Rev. Letts. *36*, 956 (1976)
144. Shand, M. L., Chance, R. R., and Silbey, R.: Chem. Phys. Letts. *64*, 448 (1979)
145. Chance, R. R., Shand, M. L., Hogg, C., and Silbey, R.: Phys. Rev. B *22*, 3540 (1980)
146. Patel, G. N.: Third Intl. Meeting on Radiation Processing, Tokyo 1980
147. Chance, R. R., Shand, M. L., LePostollec, M., and Schott, M.: J. Polym. Sci. Polym. Letts. Ed. *19*, 529 (1981)
148. Shand, M. L., Chance, R. R., LePostollec, M., and Schott, M.: Phys. Rev. B *25*, 4431 (1982)
149. Miller, G. G. and Patel, G. N.: J. Appl. Polymer Sci. *24*, 883 (1979)
150. Mann, B. and Kuhn, H.: J. Appl. Phys. *42*, 4398 (1971)
151. Clark, D. T., Fak, Y. C. T., and Roberts, G. G.: J. Electron Spectroscopy and Rel. Phenomena *22*, 173 (1981)
152. Kan, K. K., Petly, M. C., and Roberts, G. G.: Proc. Intl. Conf. on the Physics of MOS Insulators, Raleigh, Pergamon Press, (1980) p. 344
153. Kan, K. K., Roberts, G. G., and Petly, M. C.: Thin Solid Films *99*, 291 (1983)
154. Seiferheld, U. and Bässler, H.: Solid State Comm. *47*, 391 (1983)
155. Bubeck, C., Nguyen Xuan, T. H., Sixl, H., Tieke, B., and Bloor, D.: Ber. Bunsenges. Phys. Chem. *87*, 1149 (1983)

H.-J. Cantow (Editor)
Received November 10, 1983

# Spectroscopy of the Intermediate States of the Solid State Polymerization Reaction in Diacetylene Crystals

Hans Sixl
Physikalisches Institut, Teil 3, Universität Stuttgart
Pfaffenwaldring 57, D-7000 Stuttgart 80, FRG

*This contribution gives a review of recent spectroscopic investigations concerning the photophysical and photochemical primary and secondary processes of the solid state polymerization reaction in diacetylene single crystals. It will be shown, that diacetylenes are an unique model system for the study of the reaction mechanism of a solid state chemical reaction which is characterized by a variety of reaction intermediates. The polymerization reaction in these crystals is of special importance, due to the resulting polymer single crystals, which exhibit extraordinary anisotropic physical properties.*

*ESR (electron spin resonance) and optical absorption spectroscopy at low temperatures were used to analyse the individual reaction steps of the optical and thermal polymerization reactions and their kinetics. The reaction steps are*

*the photoinitiation,*
*the chain propagation and*
*chain termination reactions.*

*It is possible to exactly identify and characterize the radical species and chain structures of the reaction intermediates, which are determined by their different reactive or unreactive chain ends. The reactive intermediates are best described by diradical (DR), asymmetric carbene (AC) and dicarbene (DC) oligomer molecules of different lengths. The respective singlet ($S = 0$), triplet ($S = 1$) or quintet ($S = 2$) states and their roles in the polymerization process are investigated in detail by solid state spectroscopy. A one-dimensional electron gas model is successfully applied to the optical absorption series of the DR and AC intermediates as well as on the different stable oligomer SO molecules obtained after final chain termination reactions.*

# 1 Introduction

## 1.1 Polymer Single Crystals

Applying conventional procedures polymerization reactions are generally performed in the fluid phase [1]. As a rule the macromolecules produced in this reaction form either glassy or chainfolded microcrystalline polymer. The original extreme anisotropy of the individual "one-dimensional" polymer filaments is lost due to the arbitrary orientations of the polymer segments in such materials. Therefore in general the mechanical and optical properties of typical polymer foils are completely isotropic. The crystallization of a polymer solution or melt fails to produce macroscopic polymer single crystals, due to the enormous length and flexibility of the polymer chains [2].

The progress of spectroscopic research concerning the electronic properties of conventional organic polymers has been severely hindered by their gross structural inhomogeneities. Typical microscopic polymer crystals distributed in bulk polycrystalline polymers grow from solution (e.g. polyethylene [2]) have chain axis dimensions of about 100 Å. Moreover defects such as dislocations, chain kinks and cross-linking restrict the lengths of the quasi one-dimensional chain Sections [3, 4]. The physical properties of polymeric materials are strongly dependent on the defect structures. The isotropic distribution of the crystallites in bulky polymers excludes the investigation of the instrinsic anisotropic properties of the individual polymer filaments. Detailed analysis has not been possible with spectroscopic methods, such as ESR- or X-ray structure analysis which are most powerful with single crystalline samples. For all these reasons there has been considerable interest in macroscopic polymer single crystals.

Nearly perfect large-dimension polymer single crystals were only recently successfully synthesized from monomers by solid state polymerization [5-14]. This method has been most successful in the preparation of polydiacetylene single crystals. The reaction mechanisms of this remarkable solid state reaction will be discussed in this contribution. A second method (which yields less perfect polymer crystals or only ordered polymer filaments) involves nearly simultaneous crystallization and polymerization of monomer molecules at liquid-solid or gas-solid interfaces [15, 16]. This method has been most successful in the preparation of polysulphur nitride $(SN)_x$ "crystals", which in contrast to the large band-gap semiconducting polydiacetylene crystals [17-21] is a polymer metal with superconductive properties below about 0.3 K [22]. The interest of both chemists and physicists in polymer crystals is based on the extraordinary anisotropic optical and electrical properties expected from the regular periodic arrangement of the linear filaments in the polymer single crystals.

## 1.2 Solid-State Polymerization

In contrast to conventional methods the polymerization of diacetylenes proceeds within the original monomer crystal which can be grown from solution. In ideal cases, for example TS-6 [bis(p-toluene sulphonate) ester of 2,4hexadiyne-1,6-diol], the polymerization reaction transforms a monomer molecular crystal to a polymer

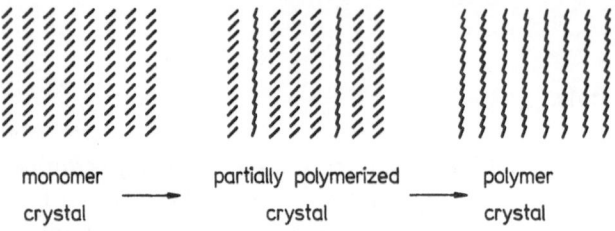

monomer ——→ partially polymerized ——→ polymer
crystal                crystal                              crystal

**Fig. 1.** Schematical representation of the solid state polymerization reaction at room temperature. The partially polymerized monomer crystal contains long polymer filaments

crystals with nearly the same dimensions and similar structural perfection [5-14]. The reaction is shown schematically in Fig. 1. Mostly the reaction is initiated photochemically [by highly energetic UV, X-ray, or γ-ray quanta (hv)] or thermally [by heat (kT), i.e., by lattice or molecular vibrations]. The polymer filaments then grow in the monomer stacking plane (see Fig. 2). In the ideal case the polymer enters the polymerizing lattice as a solid solution over the entire monomer-to-polymer conversion range. In this way phase separation between monomer and polymer with the associated shear and volume strains are avoided [23, 24]. Brillouin scattering experiments [24], scanning electron microscopy [25], gel permeation chromatography [26], optical spectroscopy [27] and laser holography [28] confirm that extremely long chains of about 100 to 1000 monomer units are formed within the partially polymerized monomer crystals.

As for most solid state reactions the "topochemical principle" [29-31] is valid in this specific polymerization reaction. It may be formulated as follows: The reactions in solids are performed with a minimum in the atomic and molecular motion (principle of least motion). Therefore, the occurance of a chemical reaction and the stereo-

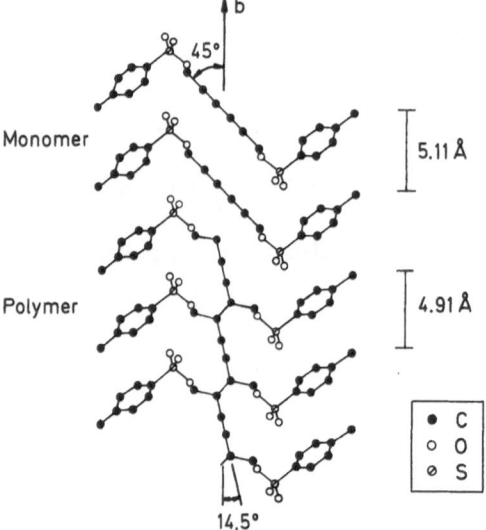

| | |
|---|---|
| ● | C |
| ○ | O |
| ⊘ | S |

**Fig. 2.** Comparison of the atomic positions in the monomer and in the polymer crystal as deduced from X-ray data [36]

chemistry of the reaction products are determined by the monomer crystal structure and the positions of the reactive groups within individual molecules. Applying these principles it is possible to synthesize highly stereoregular extended-chain polymers, which are impossible to be prepared by alternative means. The conception, therefore, is to produce crystals in which the monomers, each with their two potential reactive centres, are oriented so that they can rotate in place to link up with their neighbours. An optimal arrangement of the monomer molecules is obtained by an appropriate choice of their substituents R. The best conditions are obtained in the absence of any linear displacement of the molecular centres. In other cases a change in the lattice parameters associated with mechanical stress diminishes the reactivity. Large stress due to inhomogeneous polymerization may lead to the destruction of the crystals as for example in the case of distyrylpyrazine [32, 33] and other $2\pi + 2\pi$ photocyclizations.

The polymerization of the diacetylene crystals is usually described by the following reaction equation

$$n[R-C\equiv C-C\equiv C-R] \xrightarrow{h\nu} \left[ \begin{array}{c} C-C\equiv C-C \\ R \end{array} \right]_n \longleftrightarrow \left[ \begin{array}{c} C=C=C=C \\ R \end{array} \right]_n \qquad (1)$$

acetylene                         butatriene

structures

Fully conjugated and fully chain-aligned polymer single crystals with planar polymer backbone [34] are obtained, which may have the alternative acetylene (ynene) or butatriene structures of Eq. (1). From our experiment we know that the acetylene structure is dominant in the polymer molecules. Up to now the best investigated diacetylene crystals are the TS-6 monomer crystals and the corresponding polymer crystals (poly TS-6). The substituents R and the notation of further diacetylene crystals discussed below are listed in Table 1.

The atomic positions and the lattice parameters of the monomer and polymer crystals have been determined by X-ray analysis [34-38]. Fig. 2 shows the projection of the monomer and polymer molecules on the plane of the polymer backbone. The positions of the substituents $R = CH_2SO_3C_6H_4CH_3$ are only slightly changed, however the central diacetylene unit $-C\equiv C-C\equiv C-$ is rotated within the plane of the backbone by an angle of about 30°.

**Table 1.** Some examples of diacetylene crystals of the structure $R-C\equiv C-C\equiv C-R$, $R_1-C\equiv C-C\equiv C-R_2$ and

$$\begin{array}{c} C\equiv C-C\equiv C \\ -R_3- \end{array}$$

| Notations | Substituents |
| --- | --- |
| TS-6 | R: $-CH_2OSO_2C_6H_4CH_3$ |
| TS-12 | R: $-(CH_2)_4OSO_2C_6H_4CH_3$ |
| FBS-6 | R: $-CH_2OSO_2C_6H_4F$ |
| TCDU | R: $-(CH_2)_4OCONHC_6H_5$ |
| TCDA | $R_1$: $-(CH_2)_9CH_3$, $R_2$: $-(CH_2)_8COOH$ |
| BPG | $R_3$: $-OCO(CH_2)_3OCO-$ |

### 1.3 Spectroscopic Investigations

The polydiacetylene crystals have striking spectroscopic properties, which have been utilized in studies of their electrical [17-21], structural [34-38] and optical [39-43] behavior. These properties are in contrast to those of the monomer crystals, which show the characteristic features of molecular single crystals. As a consequence of the construction of a fully chain aligned polymer structure the mechanical and optical properties of the diacetylene crystals change dramatically during the phase transition from monomer to polymer. This effect is demonstrated in Fig. 3 by example of the optical absorption spectrum of a typical diacetylene crystal at room temperature [44]. The clear and fully transparent monomer crystals (with absorptions in the ultraviolet spectral region, starting at about 300 nm, viz. 33,000 cm$^{-1}$) become coloured with increasing polymerization from pink to deep red and finally obtain a metallic lustre in the fully polymerized state.

A direct measure of the optical absorption coefficient $\alpha$ is the optical density OD defined by $OD = \log I_0/I = 0{,}434\alpha d$ ($I_0$ and I are the incident and the transmitted light intensities and d is the thickness of the crystal). Weak polymer absorption in the range from 600 to 400 nm is present in the original monomer crystals due to weak thermal polymerization reactions. The absorption of the linear polymer molecules (which are homogeneously distributed within the partially polymerized monomer diacetylene crystals) increase during UV-irradiation at room temperature due to photopolymerization reactions. In contrast to the monomer absorption, the polymer

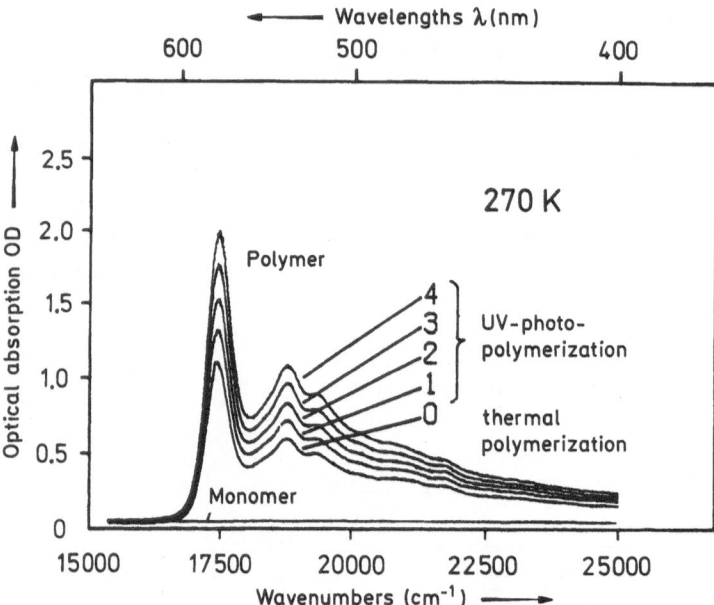

**Fig. 3.** Optical absorption spectra of a diacetylene crystal (TS-6) at about room temperature. The optical density is proportional to the absorption coefficient. The absorption in the visible spectral region is due to the polymer filaments. UV photopolymerization has been performed in steps using the same photon numbers, respectively

absorption is extremely anisotropic with $\alpha_{\parallel}:\alpha_{\perp} \approx 10^4$ [41,43]. All optical absorption spectra presented in this review are polarized with electric field vector $E$ of the incident light parallel to the chain direction $b(E\|b)$, since for $E \perp b$ no absorption of the polymer chains is observed under the prevailing experimental conditions.

At the ends of the polymer chains and at the ends of the short oligomer units (see for example the trimer molecule of Table 1) a bond defect structure is expected. For the acetylene structure of the polymer chain this is a carbene $-\overset{..}{C}-$ with two free valence electrons and in the case of the butatriene structure this is a *radical carbon* atom $-\overset{.}{C}=$ with one free valence electron. In both cases there is a reactive chain end, which allows reaction of the chain with the neighbouring monomer molecules. These reactive structures and a possible nonreactive structure are listed in Table 1 as examples of the trimer molecules.

Structure and kinetics of the radical ends (with electron spin $S = 1/2$) and of the carbene ends (with $S = 1$) are in principle investigated most effectively by the methods of electron spin resonance (ESR) and by electron nuclear double resonance (ENDOR). Information on the lengths and structures of the butatriene or acetylene short oligomer molecules should be obtained from optical absorption spectra, completely analogous to that for the well known linear dye molecules of different lengths described by the theory of Kuhn [45,46]. Both methods are highly suitable for investigating the electronic structure of the reaction intermediates and the kinetics of the solid state polymerization reaction. Due to the fast polymerization reaction at room temperature the spectroscopic methods were mainly restricted to the investigation of the polymer species. Through these studies it was, in fact, possible to detect triplet carbenes and doublet radical species at low concentrations [47-50]. However, it proved impossible to spectroscopically detect short dimer-, trimer-, . . . and oligomer-intermediates at room temperature. This is due to their extremely short lifetimes which, as we will see later, range in the nanosecond to microsecond time scale (see for example the experiments of Niederwald et al. [51,52]).

## 1.4 Reaction Mechanism

For a detailed investigation of the individual steps of the polymerization reaction, including the different intermediates of both the initiation reaction (formation of the dimer) as well as the subsequent addition polymerization reactions (formation of the trimer, tetramer, . . .), it is necessary to slow down or even stop the reaction by cooling the crystals. Consequently the concept of the following spectroscopic investigations is to photochemically initiate the polymerization reaction at extremely low temperatures (10 K) and to investigate the structure and kinetics of the intermediates obtained in subsequent photochemical or thermal reaction steps. The first low temperature experiments have been performed by Hori and Kispert using X-irradiation [53,54], and by Bubeck et al. [55,56] and Hersel et al. [57,58] using UV-irradiation.

The reaction mechanisms of this extraordinary solid state reaction has strongly attracted the interest of the polymer chemists from the very beginning. Valuable information on several aspects of the thermal polymerization reaction mechanisms, have been obtained from X-ray structural and physico-chemical investigations, discussed by Wegner [59,60], Baughman [10,11] and Chance [61]. However, the identification and detailed characterization of the individual reaction intermediates and their kinetics has been obtained only by spectroscopic investigations at low temperatures [55–58, 62–69] as discussed in this review.

From spectroscopic data, presented in the following, we conclude that the mechanism of polymerization is described by three series of intermediate states differing by the number of reactive radical or carbene chain ends: these are the *diradicals* "DR", the *dicarbenes* "DC", and the *asymmetric carbenes* "AC". Via a final chain termination reaction an additional series of reaction products is obtained. These are the *stable oligomers* "SO" with two unreactive chain ends. The schematic structures of the DR, DC, AC, and SO molecules are shown by example of the trimer in Table 2. The lengths of the dimer-, trimer-, tetramer- ... units are characterized by the numbers n = 2, 3, 4, ... of the respective monomer molecules. The symbols and the schematic structures as well as the notation of the optical and the ESR absorption lines, are summarized in Table 2.

In all experiments described in this work only extremely low concentrations of intermediates are considered. This is due to our interest which is primarily focussed on the most important initial steps of the polymerization reaction, which are characteristic of the overall polymerization reaction mechanism. Consequently only low final polymer conversion is expected and, therefore, complications arising from the interaction between the intermediate oligomer states can be neglected. It will be shown that the low temperature conventional optical absorption and ESR spectroscopy are powerful spectroscopic methods which yield a wealth of information concerning structural and dynamical aspects of the intermediate states in the photopolymerization reaction of diacetylene crystals. Therefore, this contribution will center on the photochemical and photophysical primary and secondary processes of this

**Table 2.** Symbols, spin multiplicities, structures and notations. The simplified structures of the reaction intermediates and of the final stable products are shown schematically by example of the trimer molecule

| Symbols | Structures | | Notations | |
|---|---|---|---|---|
| Spin S | Simplified structures | Chains | Optics | ESR |
| —DR— diradicals $S = 0, S^* = 1$ | | butatriene $n \geq 2$ | A, B, C, ... | I, II, III, ... |
| —DC— dicarbenes $S = 0, S^* = 1, 2$ | | acetylene $n \geq 7$ | — | i, ii, iii, ... |
| —AC— asymmetric carbenes $S = 1$ | | acetylene $n \geq 2$ | a, b, c, ... | 1, 2, 3, ... |
| —SO— stable oligomer $S = 0$ | | acetylene $n \geq 3$ | $\beta, \gamma, \delta, ...$ | — |

extraordinary solid state photoreaction. A classical partition of the individual reactions within the polymerization process into three different reaction types is possible. These are the chain initiation, the chain propagation, and finally the chain termination reactions.

# 2 The Intermediates of the Low Temperature Photoreaction

## 2.1 Optical Absorption Spectra

Optical absorption spectroscopy in the temperature regime between about 10 K and 300 K is usually performed using conventional optical absorption spectrometers (Cary, Phillips) and temperature variable helium gas flow cryostats. In order to initiate the polymerization reaction the diacetylene monomer crystals were excited into the $S_0 \rightarrow S_1$ absorption edge at about 310 nm using either a xenon high pressure arc or the 308 nm line of an excimer laser (ArF).

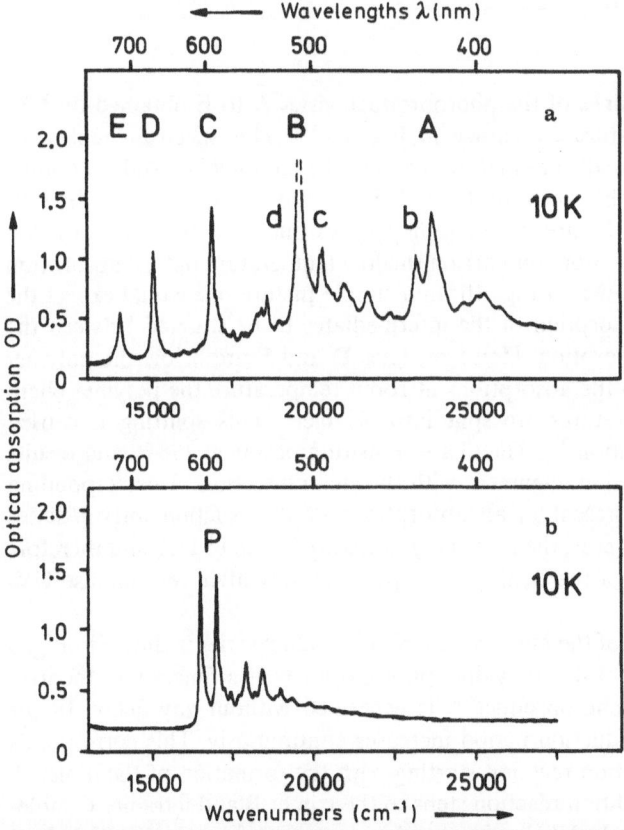

Fig. 4a and b. Optical absorption spectra of a diacetylene crystal (TS-6) at low temperatures. a Spectrum of the low temperature photoproducts obtained after UV-irradiation; b Spectrum of the polymer filaments obtained after thermal polymerization at 300 K at low degree of conversion (about 100 ppm)

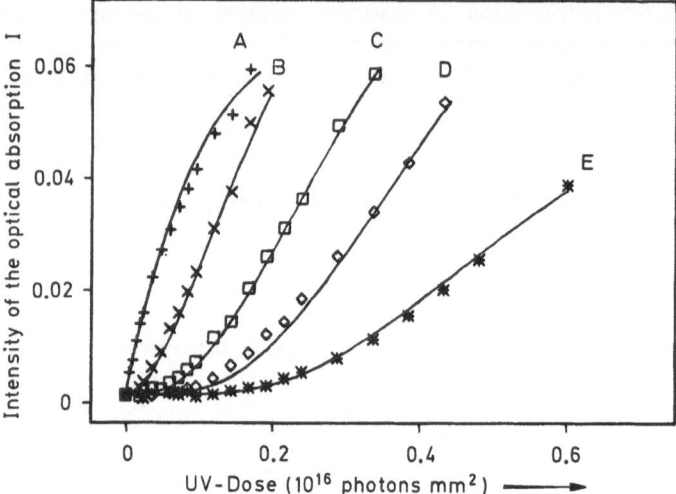

**Fig. 5.** Generation of the intermediates A to E obtained in the UV-photopolymerization reaction at 10 K. The intensity of the absorption lines is a function of the absorbed photons (proportional to the irradiation time). The calculated dependencies are fitted to the experimental points

The intense absorption lines of the photoproduct series A to E obtained by UV-irradiation at low temperature are shown in Fig. 4a [44]. This spectrum represents a difference spectrum, where the original spectrum of the monomer crystal, certaining a small amount of polymer has been substracted. Therefore only the effect of the UV-irradiation is shown in the Figure. In the same spectra lines b, c and d of a weaker series a to e are also present. For comparison the low temperature optical absorption of the polymer chains is shown in Fig. 4b. In a simple picture one would expect the positions of the optical absorption of the intermediates to be situated between the monomer and polymer absorption. However, lines D and E are below the polymer absorptions. In contrast to the absorptions at room temperature the polymer chain absorptions at low temperatures are split into doublets. This splitting is caused by a structural phase transition [35]. The phase transition occurs at 170 K and results in a doubling of the unit cell in alignment with the chain direction. A corresponding doublet structure is also present in all absorptions of the reaction intermediates described in this paper, however, their intensity ratios are less than 1:10 and therefore in most cases a resolution of the weak lines is possible only after very intense UV-irradiation.

The generation sequence of the most intense photoproduct series is shown in Fig. 5 by the integral absorptions of the individual photoproducts as a function of the irradiation time [44]. Only the photoproduct A is generated without any delay. In the sequence B, C, D, ... the induction period increases continuously. This corresponds to the expected polymerization reaction starting with the formation of the dimer A followed by subsequent addition reaction steps to the trimer B and tetramer C molecules etc. The curves are calculated using the kinetic expressions described below.

The series b, c, d, ... may be obtained in the same way by UV-irradiation (see Fig. 4a) but also in a transformation reaction by resonant optical bleaching of the series A,

B, C, ... at low temperatures. This transformation reaction is shown by example of the photoproducts A and b in Fig. 6. Upon 425 nm irradiation into the photoproduct A absorption of the A line is bleached and photoproduct b appears at the expense of A. The corresponding reactions are valid for B and c, C and d, etc.

Analogous to the photoreactivity of the A, B, C — series the b, c, d-series may be optically bleached. As a result a new series $\gamma$, $\delta$, $\epsilon$ appears in the optical absorption spectra. This series is optically and thermally stable and therefore reacts neither upon UV-or resonant irradiation nor upon annealing of the crystal up to room temperature. This is in contrast to the intermediates A, B, C, ... and b, c, d, ... which finally form polymer chains.

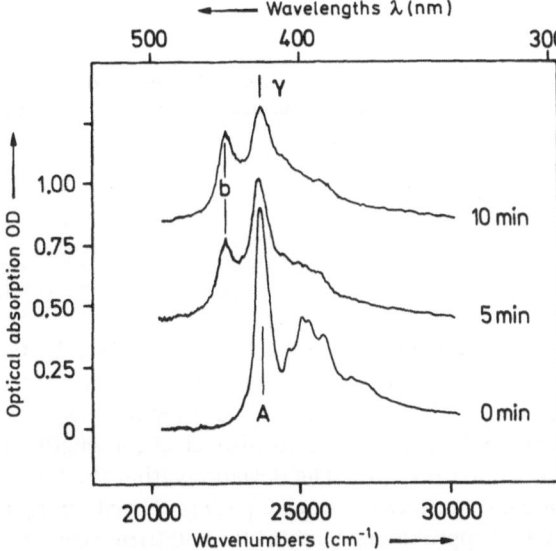

**Fig. 6.** Optical bleaching of the photoproduct A by resonant irradiation into the A absorption. The A absorption is transformed into the b and $\gamma$ absorptions (A → b → γ)

The absorption lines of the low temperature photoreaction products in TS-6 monomer crystals are summarized in the diagram of Fig. 7. The correlation of the A, B, C, ... photoproduct series to diradical DR intermediates and of the b, c, d, ... photoproducts to asymmetric carbene AC intermediates is based on the ESR experiments discussed below. The correlation of the $\gamma$, $\delta$, $\epsilon$, ... series to stable oligomers SO is based on their thermal and optical stability. The correlation of dimer, trimer, tetramer, ... molecules follows from the chemical reaction sequences observed in the time resolved optical and ESR measurements as well as from the widths of the one-dimensional potential wells used in the simple "electron gas" theory [68, 69], which already has proved successful in its application to dye molecules [45]. Following Exarhos et al. [46] the explicit dependence is given by

$$E_n = \frac{h^2}{8ml^2}(4n + 1) + E_\infty\left(1 - \frac{1}{4n}\right)$$

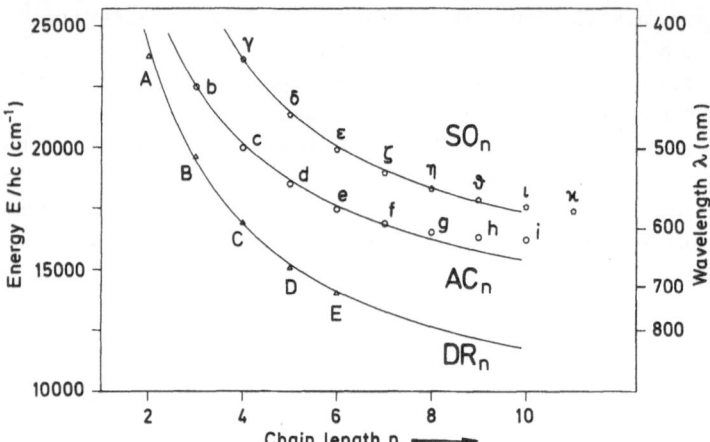

**Fig. 7.** Spectral position of the principal zero phonon absorption lines of the three photoproduct series as a function of the chain length n in monomer units. The calculated curves have been fitted to the experimental points. The data are given in Table 3

with

$$l = a_1 n + a_2 \tag{2}$$

h and m are Plancks constant and the electron mass, respectively. $a_1$ is given by the fixed lengths of the repetition unit and $a_2$ describes the boundary conditions at the ends of the oligomers. The excellent fit [68, 69] obtained by Eq. (2) supports the correlation of the different DR, AC and SO intermediates to distinct chain lengths n. There are neither excess nor missing intermediates. The deviations from the theory at long AC and SO chain length are not unexpected, since Eq. (2) is derived for equidistant carbon-carbon separations, best approximated only by the butatriene structure. The values used in the calculation are given in Table 3.

The convergence energies of the AC and SO series (see Table 3) are closest to the absorption energy of the polymer chain ($E_p \cong 17.000$ cm$^{-1}$) and consequently are suspected of having related structures. However, the convergence of the DR series A, B, C, ... is strongly red shifted in comparison with the absorption of the polymer. The diradicals DR shown in Table 1 differ essentially from the other intermediates by their butatriene structure. Consequently, the modulation of the potential curves given by the atomic positions is changed. In accordance with the theory of Brédas

**Table 3.** Data used in the one-dimensional electron gas model calculation following Eq. (2)

|            | DR            | AC            | SO             |
|------------|---------------|---------------|----------------|
| $a_1$      | 5,39 Å        | 5,39 Å        | 5,39 Å         |
| $a_2$      | 1,80 Å        | 2,10 Å        | 0              |
| $E_\infty$ | 6.600 cm$^{-1}$ | 9.700 cm$^{-1}$ | 11.200 cm$^{-1}$ |

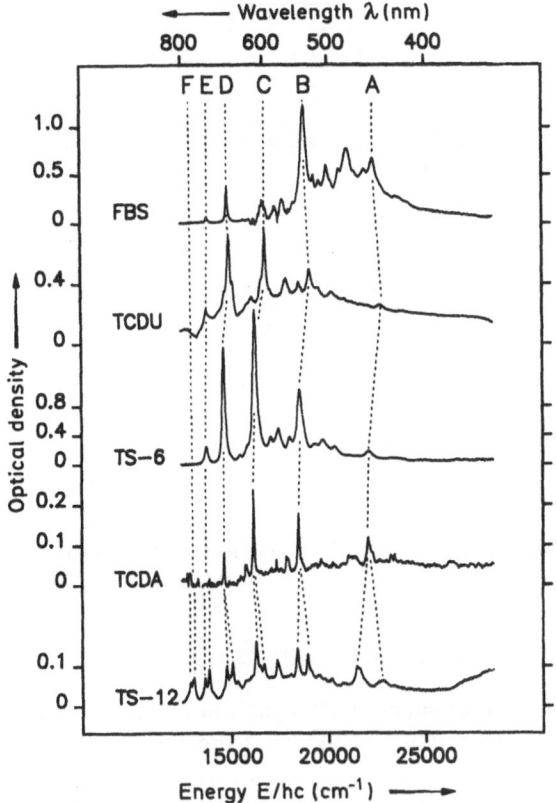

**Fig. 8.** Low temperature photoproduct absorption in diacetylene crystals with different substituents

et al. [70], distinctly lower values of $E_\infty$ for the butatriene structure of the DR series are expected as compared to the acetylene structure of the AC and SO series! The value of the absorption energy of the polymer chains in TS-6 crystals is in accordance with the proposed acetylene structure of the long polymer chains, as deduced from the original X-ray structure data [34, 71]. As will be shown later, the butatriene structure is favoured energetically as compared to the acetylene structure only at short chain lengths (photoproducts A to E) since no further longer butatrienic photoproducts F, G, H, ... could be detected in TS-6 crystals. As shown in Fig. 8 the DR- and AC-absorption series of the short oligomer units in TS-6 crystals are present also in the low temperature photoreactions in TS-12, TCDU, and TCDA crystals at almost the same spectral positions [69], though the final positions of the polymer absorptions in these crystals are very different.

## 2.2 ESR Absorption Spectra

The ESR experiments are performed in an analogous manner to the optical experiments with the same low temperature and UV irradiation equipment using a conventional ESR spectrometer (9,3 GHz, Varian).

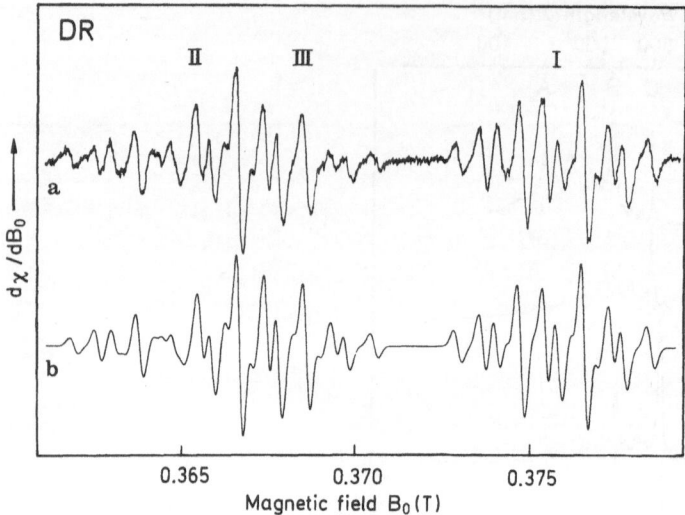

**Fig. 9a and b.** High field part of the triplet ESR spectrum of the diradical series (77 K). **a** experimental spectra; **b** simulated spectra. The correct hyperfine pattern is obtained by the interaction with two $CH_2$-groups

The high-field ESR spectrum of the most intense ESR signals of the diradical DR series I, II, III obtained after UV-photoinitiation of the low temperature polymerization reaction is shown in Fig. 9a. This spectrum is part of a typical triplet ESR spectrum with spin quantum number $S = 1$ which consists of two ESR lines with $\Delta m = \pm 1$. The splitting of the ESR line pairs (the low-field lines are not shown in the Figures, see e.g. Ref. [56]) is the fine-structure splitting due to the magnetic dipolar interaction of the two electron spins, which are electrostatically coupled to form either triplet states or singlet states. Due to the doubling of the unit cell in the low temperature phase of the TS-6 crystals the ESR spectra show a further line splitting (analogous to the splitting of the polymer absorption in Fig. 4b) which is not relevant for our purposes.

The temperature dependence of the individual ESR lines is shown in Fig. 10. The reversible disappearance of the lines at low temperatures clearly demonstrates a thermal activation of the triplet species. Optimum signal intensities are obtained at about 77 K (see Fig. 9a). The temperature dependence of the individual lines is best described by the energy level system shown in Fig. 10 with a singlet $S = 0$ ground state and an activated triplet state. The corresponding intensity dependencies are given by $I(T) \propto x/(3 + e^x)$ with $x = E_T/kT$ (k is Boltzmann's constant) shown by the curves in Fig. 10. The activation energies $E_T$ of the triplet states are obtained from the optimal fit of the curves. They are different for the individual triplet states and lie between 7 and 14 meV [56]. The temperature dependence of the ESR signal intensities is reversible only below 80 K. At higher temperatures the ESR signals disappear irreversibly due to the thermal reactions which finally lead to polymer filaments.

The hyperfine interactions arise from the magnetic dipolar coupling of the triplet electrons with the protons (nuclear spin $I = 1/2$) within the $CH_2$-groups of the sub-

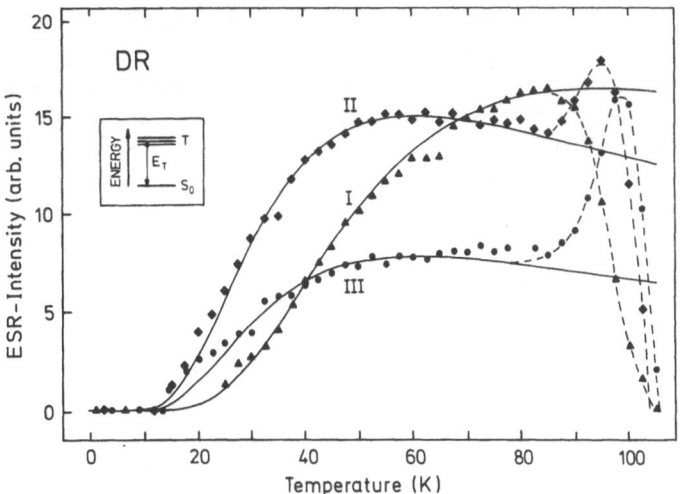

**Fig. 10.** Temperature dependence of the diradical spectra. The calculated temperature dependencies using the inserted energy level diagram has been fitted to the experimental points

stituents $R = CH_2R'$ which link the substituents to the diacetylene unit. The hyperfine splitting of the ESR lines has been analyzed by computer simulation. The calculated spectra of Fig. 9b are based on a hyperfine interaction with only two identical $CH_2$-groups located at the two ends of the oligomer units.

Owing to the low values of their fine structure parameters and to their characteristic hyperfine pattern the thermally activated triplet states I, II, III, ... are identified as diradical DR oligomers with the following symmetric butatriene structure:

$$\begin{array}{c} \text{(structure)} \end{array} \qquad DR_n \qquad (3)$$

with n = 2, 3, 4 ...

This structure is in accordance with the optical absorption spectra, where a sequence of butatriene chain structures with increasing n was postulated as responsible for the red shift of the absorption lines A to E (see Fig. 7). The longest diradical molecule (which has been detected in the optical system) therefore is the hexamer E with n = 6. For each value of n further mesomeric butatriene and acetylene structures are present up to maximally 10% of the real structure of the diradicals.

By analogy to the optical absorptions A to E (see Fig. 5) the kinetical generation sequence of the diradical oligomers I to III obtained upon UV-irradiation of the monomer crystals is shown in Fig. 11 [62]. Only the ESR line I is formed without any delay. The curves are calculated using the kinetic model described below. In this model the ESR line I corresponds to the DR dimer molecule, II to the DR trimer molecule, etc.

Parallel to the generation of the DR oligomers carbenoid AC oligomers (see Table 2) are formed upon UV-irradiation of the monomer crystals at low temperatures. The high field part of the AC triplet ESR spectra with the lines 1, 2, 3, and 4 is shown in Fig. 12. The unspecified lines are not relevant for our purposes; they are attributed to the doubling of the unit cell in the low temperature phase of the TS-6 crystals [35].

In contrast to the DR triplets, the AC triplets are in their electronic ground state. This follows from the $1/T$ Curie law temperature dependence of the ESR intensities. The fine structure splitting and the pattern of the hyperfine structure of the DR and

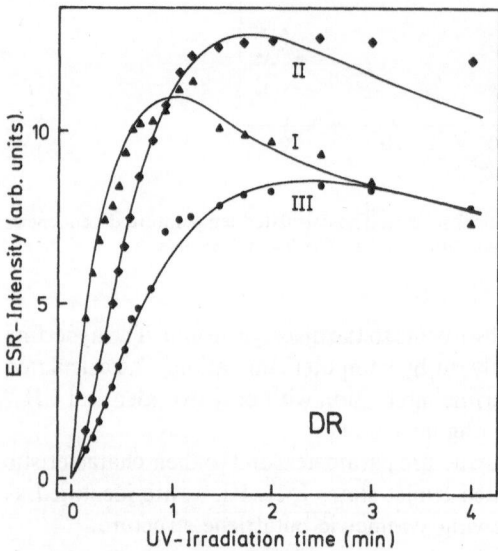

**Fig. 11.** Diradical formation upon UV-irradiation. The calculated dependencies have been fitted to the experimental points

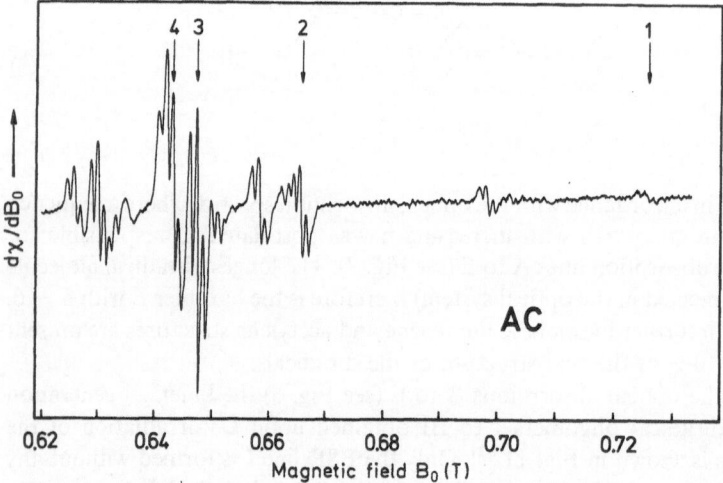

**Fig. 12.** High field part of the triplet ESR spectrum of the asymmetric carbenes. Only the high field partners of the pair lines have been assigned by the numbers

AC triplet states are totally different. The large fine structure splitting of the AC triplets is reflected in the high values of the resonance fields in Fig. 12. The corresponding large fine structure parameters are characteristic of carbenoid states [72-76].

The intensity ratio of 1:2:1 of the three hyperfine components of one AC-ESR line follows from the interaction of the triplet electrons with the $CH_2$-rest group of only one chain end. From the fine structure and hyperfine analysis [55] the total carbene and butatriene contributions are estimated to be 60% and 40%, respectively.

The principal mesomeric structures of the AC- end group are given by

$$R-\overset{\bullet}{\underset{\bullet}{C}}-C\equiv C-C\overset{\parallel}{\underset{R}{\diagdown}}\cdots \longleftrightarrow R-C\equiv C-\overset{\bullet}{\underset{\bullet}{C}}-C\overset{\parallel}{\underset{R}{\diagdown}}\cdots \longleftrightarrow R-\overset{\bullet}{C}=C=\overset{\bullet}{C}-C\overset{\parallel}{\underset{R}{\diagdown}}\cdots \tag{4}$$

The structure of the AC molecules therefore is best represented by

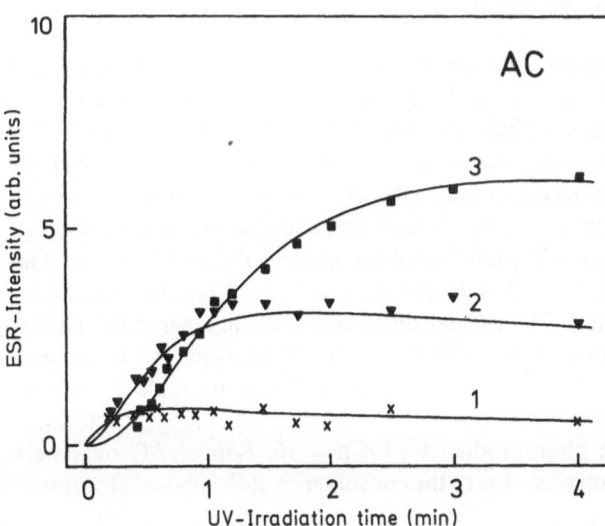

$$ AC_n \tag{5}$$

with n = 2,3,4...

The structure of the unreactive chain end will be discussed below.

The dominant structure of the oligomer chain attached to the carbene end is the acetylene structure $\left(RC-C\equiv C-CR\right)_{n-2}$. As discussed above this structure is also responsible for the spectral positions of the optical absorption lines b, c, d, ... shown in Fig. 7.

The kinetical generation sequence of the AC ESR signals upon UV-irradiation of the monomer crystal is shown in Fig. 13. Only the ESR line 1 is formed without

Fig. 13. Formation of the asymmetric carbenes upon UV-irradiation. The calculated dependencies have been fitted to the experimental points

any delay. The curves are calculated using the kinetic model described below. In this model the ESR line 1 corresponds to the AC dimer molecule, 2 to the AC trimer molecule etc.

Besides the direct generation of the monomer crystals by UV-irradiation AC molecules may also be produced indirectly by optical bleaching of the DR molecules. This has been demonstrated by example of the optical absorption A → b in Fig. 6. The corresponding transformation of the ESR lines I → 2 during optical bleaching of A → b is shown in Fig. 14. The dimer DR molecule I is *not* transformed to the dimer AC molecule 1 but to the trimer AC molecule 2. Consequently, one monomer molecule is added in each resonant phototransformation reaction of DR molecules to AC molecules.

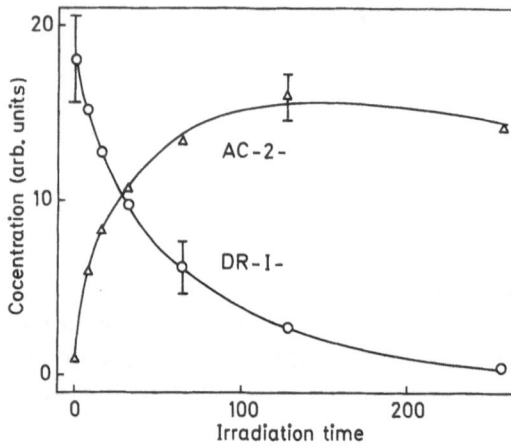

Fig. 14. Transformation of the diradical ESR line I to the asymmetric carbene ESR line 2 upon resonant irradiation into the photoproduct A absorption. The concentration of the triplet species has been deduced from the evaluation of the ESR spectra

## 2.3 Correlation of the Photoproducts

The resonant phototransformation reaction allows a clear correlation of the optical and ESR spectra: The series A, B, C, ... and I, II, III correpond to the same diradical DR photoproducts with butatriene chain structure. The series a, b, c, ... and 1, 2, 3 correspond to the same asymmetric carbene AC photoproducts with acetylene chain structure. The identification of the dimer molecules $DR_2 = A = I$ and $AC_2 = a = 1$ is based on their spectral positions, on the generation, addition and transformation properties, and on their quadratic photoinitiation process discussed below. The identification of the trimer ($DR_3 = B = II$ and $AC_3 = b = 2$) and tetramer molecules, etc. follows the same lines. The longest detectable DR oligomer is the photoproduct E which is the hexamer molecule with n = 6. In TS-12 crystals [27] the longest DR oligomer is the heptamer with n = 7. Longer DR units are unstable. As will be shown below they are transformed to dicarbene DC molecules. The longest detectable AC oligomer molecule is the photoproduct i with n = 10. Longer AC oligomers probably occur but are not resolvable, due to the convergence of the optical absorption spectra.

The asymmetric AC species are composed of a reactive carbenoic chain head and an unreactive chain tail with fully saturated bonds. They are formed by optical transformation of the diradical molecules $DR_n$: In a first chain termination reaction one reactive chain end is destroyed and an unreactive chain end is formed by addition of a monomer molecule following $DR_n \rightarrow AC_{n+1}$. In a second chain termination reaction the remaining reactive chain end of an AC oligomer may be destroyed in a further photoaddition reaction. In this way the AC oligomers are transformed to stable oligomers SO ($\gamma$, $\delta$, $\varepsilon$, ...) with two unreactive chain ends. This transformation reaction has been observed in the optical spectra. As expected no corresponding ESR spectra could be detected since radical electrons were absent.

The correlation of the stable oligomers observed in the optical spectra (see Fig. 7) to trimer ($\beta = SO_3$), tetramer ($\gamma = SO_4$), pentamer ($\delta = SO_5$) molecules etc. is based on the $AC_n \rightarrow SO_{n+1}$ transformation process. From the convergence of the optical absorption lines (see Fig. 7) an acetylene structure of the SO backbone similar to that of the AC oligomers is very likely.

The structure of the SO molecules therefore is best represented by

$$
\text{R'HC} = \overset{|}{\underset{H}{C}} - C \equiv C - \overset{R}{\underset{R}{C}} \underset{n-2}{\overset{R}{\underset{R}{C}}} - C \equiv C - \overset{R}{\underset{R}{C}} \cdots \overset{R}{\underset{R}{C}} - C \equiv C - \overset{H}{\underset{CHR'}{C}} \qquad SO_n
\tag{6}
$$

with n = 3,4,...

So far the structure of the unreactive SO chain ends has not been clearly proved.

Three essentially different stable chain end configurations have been suggested. The following mesomeric planar pseudocyclopropen structures (7) with bond length up to 1,6 Å have been suggested by Wegner [58].

$$
\tag{7}
$$

The following nonplanar methylene-group binding, which is formed by an inter-molecular insertion reaction of the reactive radical chain end into a CH-bond of the rest group $R = CH_2R'$ of an adjacent monomer molecule has been suggested by Bubeck [77].

$$(8a)$$

The unreactive chain end structures of the $AC_n$ and $SO_n$ molecules shown in Eqs. (5) and (6) are obtained by an intramolecular insertion reaction given by

$$... -C \equiv C - \overset{.}{C} - CH_2 - R' \rightarrow ... -C \equiv C - CH = CH - R'. \qquad (8b)$$

All structures (7) and (8) are hypothetical, because up to now no final proof of the structures — for example by analysis of the vibronic structures or by chemical shift measurements — has been given. The reactions leading to the discussed structures (7) and (8) are typical carbene reactions [78, 79].

## 2.4 The Individual Steps of the Low Temperature Photopolymerization Reactions

Following the experimental absorption data a classical partition of the low temperature photopolymerization reaction is possible. The individual reaction steps are given by the photoinitiation reaction (formation of reactive dimers), the photoinduced chain propagation (addition of monomer molecules to the reactive oligomers), and the photoinduced chain termination reaction (formation of unreactive chain ends). As shown in the optical and ESR experiments the dimer $DR_2$ and $AC_2$ formation is quadratically dependent on the UV-light intensity [62, 63], whereas both the chain propagation and the chain termination reactions are linearly dependent on the intensity. Therefore, the dimer photoinitiation is a bimolecular reaction in contrast to the monomolecular photoaddition reactions, leading to chain propagation or chain termination.

In the bimolecular chain initiation reaction two monomer molecules must first be photoexcited according to $2M + 2h\nu \rightarrow 2M^*$. Then via energy transfer ET processes the excitation energy (which may be of excitonic, vibronic, libronic, or phononic character) must be transferred to the immediate neighbourhood of the reaction center. As will be shown later (see Section 4) the reactive centre has a radical structure (which is obtained by electronic excitation of the monomer molecules $M^*$) whereas the second excitation encountering the adjacent molecules $M'$ of the reaction centre need not necessarily be of radical nature. The energy transfer is described by $M^*M' + M^* \rightarrow M^*M'^* + M$. From the ESR experiments the reaction yields of the $DR_2$ and $AC_2$ dimer molecules are given by approximately 10:1. The low concentration of the $AC_2$ centres (denoted by 1) and the energetic position of the corresponding optical absorptions close to that of the monomer explains their apparent absence from the optical spectra.

The photoinitiation reactions are described by the following reaction Equation:

$$2M + 2h\nu \longrightarrow 2M^* \xrightarrow{\text{ET}} M^*M'^* \begin{array}{c} \nearrow^{k_1^{DR}} DR_2 \\ \searrow_{k_1^{AC}} AC_2 \end{array} \tag{9}$$

The photoinitiation reactions is possible only upon excitation of the monomer molecules M* of the monomer crystal. No further dimer formation is possible when the crystal is irradiated below the monomer absorption band located at 310 nm.

Parallel to the photoinitiation processes (with hν) photoaddition processes are observed, as shown for example in the Figs. 4, 5, 11 to 13. After dimer initiation, trimer formation from the dimer is possible etc. The chain propagation within the DR or AC series is performed by photoaddition of monomer molecules M' adjacent to the reaction centres, given by the dimer ($DR_2$ or $AC_2$), trimer ($DR_3$ or $AC_3$), ... molecules. The molecules M' are lowered in energy by the perturbation introduced by the reaction centres. They form a trap for the optical excitation energy. They may be excited directly (M' + hv' → M'*) or indirectly via nonperturbed monomer molecules (M + hv → M*) and subsequent energy transfer (M* + M' $\xrightarrow{\text{ET}}$ M + M'*). The chain propagation reaction therefore is in competition with the chain initiation reaction.

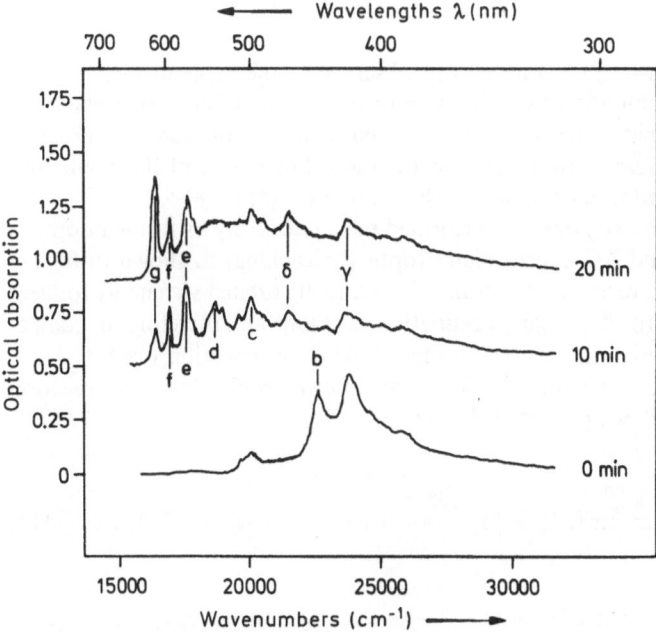

**Fig. 15.** Optical addition polymerization of the AC series b to g upon 364 nm irradiation of the TS-6 crystal. The initial spectrum at 0 min corresponds to the final spectrum of Fig. 6. In addition transformation reactions to the γ, δ-series are observed

Direct excitation of the molecules M′ is performed with UV-light energies below the monomer absorption edge hv′ < hv (with wavelength 320 nm < λ < 370 nm). The photoaddition polymerization reaction obtained in this way at 10 K is shown for the optical absorptions of the AC-centres (b, c, d, ...) in Fig. 15. After preparation of the b photoproduct (see the procedure of Fig. 6) the sequence of the AC centres b → c → d → e → f → g is obtained upon 364 nm-irradiation. Simultaneously chain termination reactions leading to the stable oligomers δ and ε are observed. The same reactions are also observed with the diradicals in the ESR and optical spectra. After preparation of the dimer diradical and subsequent 364 nm-irradiation, very effective addition polymerization reactions are observed parallel with chain termination reactions leading to AC-molecules. The same effects are valid for the dicarbene DC species described below (Section 3.2).

The photoinduced chain propagation by addition of adjacent monomer molecules M′ to the DR, AC and DC reaction centres is described by the following reaction Equations:

$$DR_n + M' \underset{}{\overset{hv'}{\rightleftharpoons}} DR_n + M'^* \xrightarrow{k_n^{DK}} DR_{n+1}; \qquad n = 2, 3, \ldots 5, \qquad (10\,a)$$

$$AC_n + M' \underset{}{\overset{hv'}{\rightleftharpoons}} AC_n + M'^* \xrightarrow{k_n^{AC}} AC_{n+1}; \qquad n \geq 2. \qquad (10\,b)$$

$$DC_n + M' \underset{}{\overset{hv'}{\rightleftharpoons}} DC_n + M'^* \xrightarrow{k_n^{DC}} DC_{n+1}; \qquad n \geq 7. \qquad (10\,c)$$

No back reactions (n → n − 1) have been observed in the experiments.

Parallel to the photoinitiation (with hv) and to the photoaddition reactions (with hv and hv′), chain termination reactions are observed as shown, for example, in Figs. 4, 6 and 15. In the chain termination reactions the radical character of the chain ends is destroyed in a photoaddition reaction as shown by Eqs. (5) to (8).

The chain termination reactions are performed most effectively by resonant optical excitation of the DR and AC intermediates (optical bleaching) as shown in Figs. 6 and 14. The DR and AC reaction centres may be indirectly excited via energy transfer from the monomer matrix. The chain termination reactions are, therefore, in competition with the photoinitiation and photoinduced chain propagation reactions.

The chain termination reactions involving an intermolecular insertion reaction are described by the following reaction Equations.

$$M' + DR_n \underset{}{\overset{hv_n^{DR}}{\rightleftharpoons}} DR_n^* + M' \xrightarrow{k_n^{DR-AC}} AC_{n+1}; \quad \text{with } n = 2, 3, \ldots 6, \quad (11\,a)$$

$$M' + DC_n \underset{}{\overset{hv_n^{DC}}{\rightleftharpoons}} DC_n^* + M' \xrightarrow{k_n^{DC-AC}} AC_{n+1}; \quad \text{with } n \geq 7, \qquad (11\,b)$$

$$M' + AC_n \underset{}{\overset{hv_n^{AC}}{\rightleftharpoons}} AC_n^* + M' \xrightarrow{k_n^{AC-SO}} SO_{n+1}; \quad \text{with } n \geq 2. \qquad (11\,c)$$

In the case of an intramolecular insertion reaction (8b) no addition of a monomer molecule is expected upon chain termination

$$DR_n \underset{}{\overset{h\nu_n^{DR}}{\rightleftharpoons}} DR_n^* \xrightarrow{k_n^{DR-AC}} AC_n; \qquad \text{with } n = 2, 3, \ldots 6, \qquad (11\,d)$$

$$DC_n \underset{}{\overset{h\nu_n^{DC}}{\rightleftharpoons}} DC_n^* \xrightarrow{k_n^{DC-AC}} AC_n; \qquad \text{with } n \geq 7, \qquad (11\,e)$$

$$AC_n \underset{}{\overset{h\nu_n^{AC}}{\rightleftharpoons}} AC_n^* \xrightarrow{k_n^{AC-SO}} SO_n; \qquad \text{with } n \geq 2. \qquad (11\,f)$$

The destruction of the first reactive chain end leads to AC oligomers. The destruction of the second reactive chain end leads to SO oligomers. No back reactions ($n \rightarrow n - 1$) have been observed in the experiments.

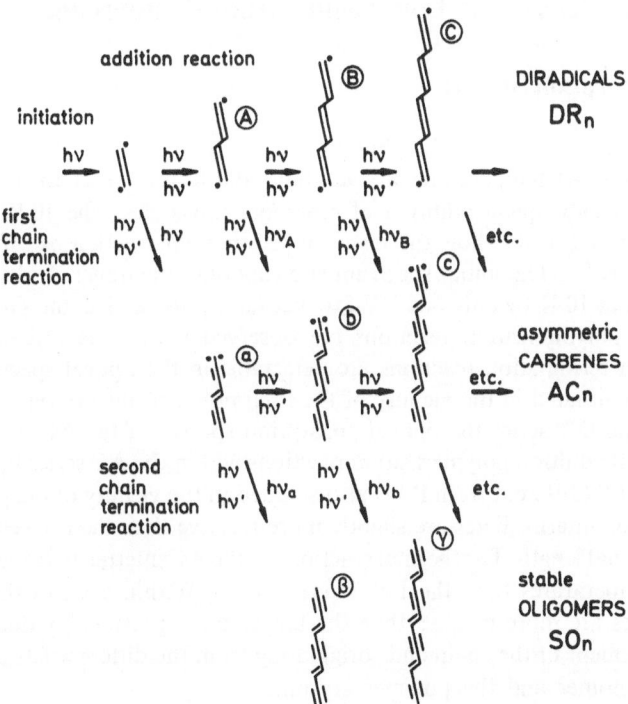

MONOMER     MONOMER     DIMER     TRIMER     TETRAMER
ground state     excited state          ground state intermediates

**Fig. 16.** Photoreactions at low temperatures. The notation corresponds to the optical absorption lines of the reaction intermediates. An intermolecular chain termination reaction is assumed. The dimer initiation reaction requires two photons ($h\nu$). The photoaddition reaction is a one photon process ($h\nu$ or $h\nu'$). The chain termination reactions are most effectively performed by resonant irradiation into the absorption of the DR or AC intermediates

### 2.5  Reaction Scheme of the Low Temperature Photoreactions

A reaction scheme showing the individual steps of the low temperature photopolymerization reaction of the DR and AC reaction intermediates and of the SO, which
are the final reaction products, are taken from the optical absorption spectra. All
molecules shown in Fig. 16 have been observed in either the optical spectra or ESR-
spectra or in both with the exception of only the short lived monomer species. The
reactive chain ends are best represented by the diradical and carbene structures,
shown in the Figure.

The photoinitiation is a two step reaction: Two photons of energy hv are involved
in the DR or AC dimer formation. Subsequent photoaddition polymerization reaction
in the DR or AC sequence are given by the horizontal arrows. They are possible
with photons of energy hv or hv'. Chain termination reactions are given by the diagonal
arrows. In these reactions the reactive chain ends are transformed to nonreactive
chain ends by resonant absorption of light quanta with energy $hv_n$ or via energy
transfer with hv or hv'. In the reaction scheme it is assumed that monomer molecules
are added in the chain termination reaction according to (11a–c).

## 3  Thermal Reactions of the Low Temperature Photoproducts

### 3.1  Optical Absorption Spectra

The DR and AC intermediates of the photopolymerization reaction are stable only
at low temperatures. At temperatures above about 100 K they react to form long
macromolecules by subsequent addition of monomer molecules. The 10 K optical
absorption spectra of Fig. 17 show the result of the thermal reaction as a function
of the time at 100 K [69]. The initial spectrum showing only the dimer A absorption
has been prepared at 10 K by only one UV-excimer laser pulse at 308 nm. Only pure
thermal addition polymerization reactions are observed within the DR-series A,
B, C, ... No chain termination reactions are detectable in the optical spectra. The
final product P' is situated in the vicinity of the final polymer absorption.

Analogous to the DR series the optical absorption spectra of the AC series also
shows pure thermal addition polymerization reactions within the AC series b, c, d, ...
The final product P'' (different from P') is also situated in the vicinity of the polymer
absorption. The AC intermediates are slightly more reactive than the respective DR
intermediates of equal length. The thermal reactions of the AC intermediates generally
start at lower temperatures than the DR intermediates. Within a series the short
oligomer molecules are more reactive than the longer ones, presumably due to the
increasing displacement of the chain ends originating from the different lattice parameters of the monomer and the polymer crystals.

The absorptions of the SO-oligomers γ, δ, ε are both, thermally and optically
stable. Therefore upon thermal annealing up to room temperature and upon UV
and resonant irradiation no reactions could be observed [69]. These oligomer molecules
can be prepared optically only at low temperatures and are the only photoproducts
which remain stable at 300 K. After preparation they may be dissolved from the

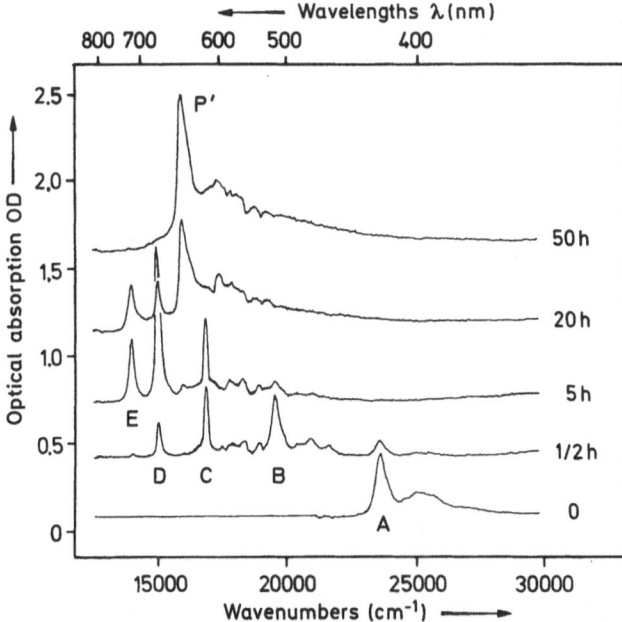

**Fig. 17.** Thermal addition reactions of the diradical series A to E at 100 K as a function of the annealing time. Optical absorption spectra

partially polymerized monomer crystal at room temperature and may be investigated in future experiments in solution with physical and chemical methods. Short trimer, tetramer, ... SO molecules cannot be prepared at high temperatures, due to the very efficient thermal addition polymerization following the photoinitiation reaction. The SO chain lengths distribution, obtained at about 300 K, is supposed to have its maximum at high chain length (n > 50), owing to the low chain termination yield in the thermal reactions.

## 3.2 ESR-Spectra

In analogy to the procedure of the optical spectra the thermal addition polymerization reaction is also observed in the ESR-spectra of the triplet DR and AC reaction intermediates [63]. In the temperature dependencies of the AC and DR centres of Fig. 18 the temperature has been increased continuously. The initial slight increase of the DR ESR signals in Fig. 18a characterizes the thermally activated DR triplet states. The chemical reactions start at about 95 K. The initial slight decrease of the AC ESR signals in Fig. 18b is given by the $1/T$ Curie law of the AC triplet ground states. The chemical reactions start at about 90 K.

The thermal reactions of the DR-intermediates, however, exhibit an extraordinary effect. Hand in hand with the disappearance of the DR series a new series appears, which shows a four line ESR fine structure arising from S = 2 quintet dicarbene DC states [63-66]. The thermal transformation of the DR intermediates into the DC inter-

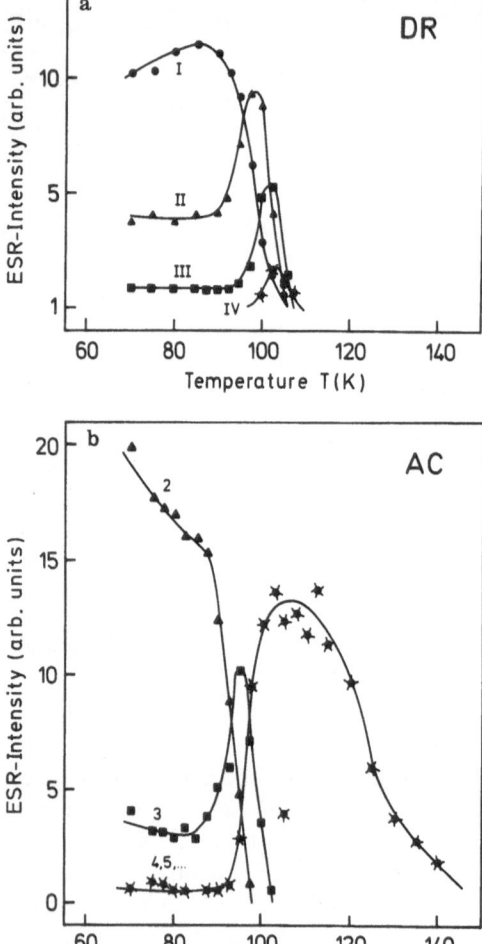

**Fig. 18a and b.** Thermal addition reactions of the diradicals (**a**) and of the asymmetric carbenes (**b**) as a function of the temperature. ESR intensities

mediates is shown in Fig. 19 as a function of the annealing time at 90 K. At 100 K the individual DC oligomers react in turn by thermal addition reaction steps to longer DR oligomers. The longest DC oligomers cannot be distinguished from the AC triplet states due to the convergence of the DC-ESR spectral lines which occurs at the same position of the long AC triplet states. In contrast to the photoinduced addition reactions no chain termination reactions could be clearly proved.

Analogous to the triplet DR-spectra the quintet DC-spectra are thermally activated. The temperature dependence of the individual quintet ESR intensities $I(T)$ could be fitted with a singlet $S = 0$ ground state and an activated quintet state. The corresponding dependencies are given by $I(T) \propto x/(5 + e^{-x})$ with $x = E_Q/kT$ [63−66]. The activation energies $E_Q$ of the quintet states (obtained from the optimal fit of the curves) are different for the individual quintet states and are extremely low in energy. As shown in Fig. 20 [67] they lie between about 30 μeV and about 10 meV. The quintet ESR intensities are reversible only below 80 K.

The fine structure and the hyperfine structure of the DC intermediates has been investigated in detail by ESR [63-66] and ENDOR spectroscopy [67]. Fig. 21 shows the angular dependence of the ESR resonance fields with $B_0$ approximately perpendicular to the oligomer chain direction. The curves represent the calculated

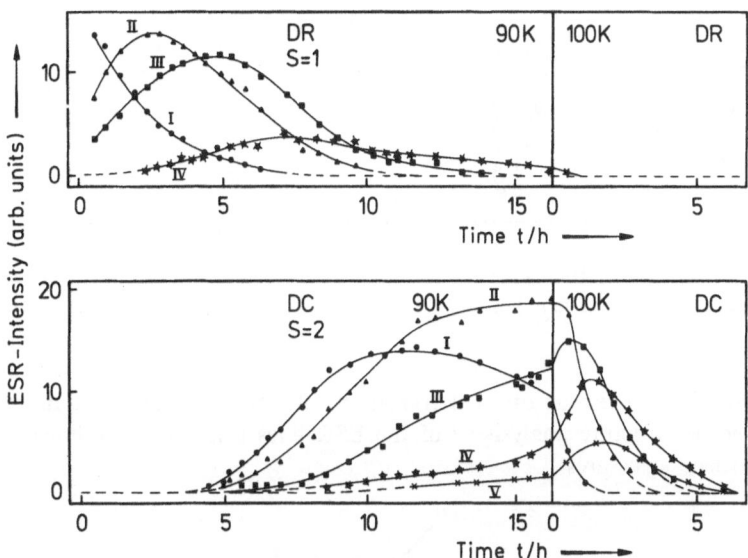

**Fig. 19.** Thermal addition and transformation reaction of the diradicals DR to the dicarbenes DC. The formation and the decay of the ESR intensities is given as a function of the time of thermal annealing at 90 K resp. 100 K

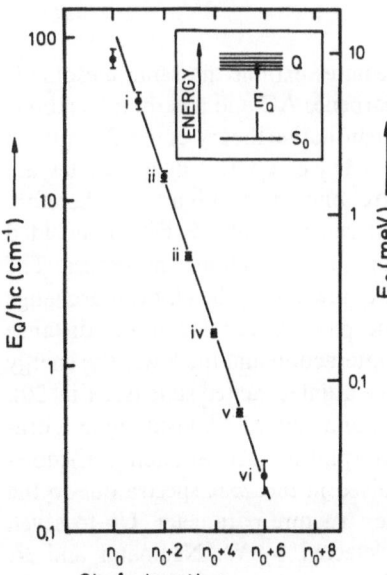

**Fig. 20.** Experimental values [67, 83] of the singlet-quintet activation energies as a function of the lengths of the dicarbene molecules. The first DC molecule presumably is the heptamer with $n_0 = 7$

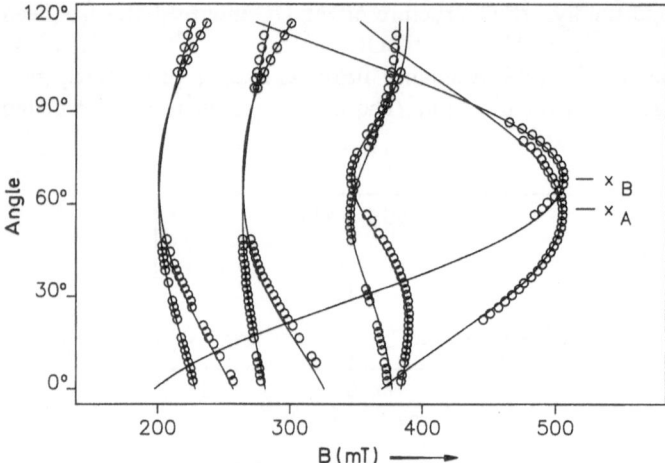

**Fig. 21.** Angular dependence of the four $\Delta m = 1$ ESR transitions of a DC-quintet state (I)

angular dependencies of the four $\Delta m = 1$ transitions of a $S = 2$ dicarbene quintet state. The hyperfine structure analysis [67] of the ESR lines leads to the following symmetric dicarbene structures:

$$\text{DC}_n \qquad n \geq 7 \tag{12}$$

The individual chain ends of the DC intermediates exhibit the same mesomeric structures (4) of the corresponding asymmetric carbenes AC and a carbon backbone with acetylene structure. It has been shown in a quantitative theory of the DC quintet states [66, 80, 81], that the fine structure parameters $D_Q$ of all DC intermediates are given by the fine structure parameters $D_T$ of two AC-oligomers with n $\approx$ 4. Therefore in accordance with the disappearance of the DR intermediates at n > 6 it is concluded that the shortest DC intermediates are either heptamer or octamer molecules. The different activation energies $E_Q$ of the quintet states are due to the electron exchange interaction between the electrons of the carbene pair [82]. The larger the distance between the carbene pair the lower the exchange interaction and the lower the energy separation between the singlet ground state and the quintet excited state (see Fig. 20). As a consequence the singlet and quintet states become mixed [80], resulting in a dramatic but continuous change of the fine structure splitting. The different DC-molecules with lengths up to 5 nm may be easily resolved in the ESR spectra due to the low value of $E_Q$ which acts as an additional fine structure parameter. Up to seven differently long DC intermediates have been detected by W. Neumann and R. Huber [66, 67]. The lengths of the DC oligomers therefore range between $7 \leq n \leq 14$

The reactions within the DR and AC series are given by the probabilities $(1 - x_n)$ and $(1 - y_n)$, respectively. Because the chemical character of the intermediates is not a function of chain length the probabilities are assumed to be independent of n, therefore $x_n = x$ and $y_n = y$. In this model, due to the branching of the photochemical reactions, the intermediates are not expected to become infinitely long. Thus, a distribution of chain lengths will be obtained for the DR intermediates, different from those of the AC and SO intermediates. The different chain length distributions — obtained under stationary conditions — are dependent on the probabilities x and y and on the different rate constants $k_n^{DR}$ and $k_n^{AC}$. They follow from the kinetic equations

$$d[DR_n]/dt = (1 - x) \, k_{n-1}^{DR}[DR_{n-1}] \, [M'^*]$$
$$- \, k_n^{DR}[DR_n] \, [M'^*] = 0 \,, \qquad n \geq 2 \,,$$

$$d[AC_n]/dt = (1 - y) \, k_{n-1}^{AC}[AC_{n-1}] \, [M'^*]$$
$$+ \, xk_{n-1}^{DR}[DR_{n-1}] \, [M'^*]$$
$$- \, k_n^{AC}[AC_n] \, [M'^*] = 0 \,, \qquad n \geq 3 \,,$$

$$d[SO_n]/dt = + \, yk_{n-1}^{AC}[AC_{n-1}] \, [M'^*] > 0 \,, \qquad n \geq 3 \,. \tag{20}$$

Stationary conditions are obtained after prolonged UV-irradiation times. In order to obtain a better understanding we will concentrate on the simplest situation characterized by $k_n^{DR} = k_n^{AC} = k$. If we refer to the first stable diradical $DR_2$ (denoted by A in the optical spectra and by I in the ESR spectra), we obtain the following distribution described by:

$$[DR_n] = (1 - x)^{n-2} \, [DR_2] \,, \qquad\qquad n \geq 2 \,,$$

$$[AC_n] = (1 - y) \, [AC_{n-1}] + x(1 - x)^{n-3} \, [DR_2]$$

with

$$[AC_2] = \frac{x[DR_2]}{1 - x} \,, \qquad\qquad n \geq 3 \,,$$

$$[SO_n] = y[AC_{n-1}] \, k \int_0^t [M'^*] \, dt \,. \tag{21}$$

In contrast to the DR and AC intermediates the concentration of the SO molecules is dependent on the UV-irradiation time. All concentrations are independent of the UV-intensity.

Figure 23 shows the stationary distribution of the DR- and AC-intermediates and of the SO molecules obtained with Eq. (21) for $x = 0.2$ and $y = 0.1$. The distribution of the DR and AC molecules has been normalized on the $DR_2$ concentration, representing the shortest DR intermediate, which are stable at low temperatures in contrast to the short-lived excited $DR_1$ molecules. The distribution function of the SO molecules has been normalized with $[SO_3] = 1$. The $SO_3$ molecules are the shortest stable oligomer molecules.

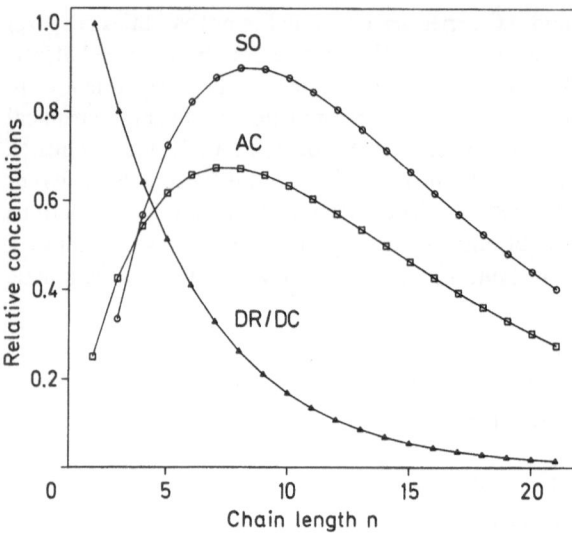

**Fig. 23.** Calculated chain length distribution in TS-6 during stationary UV-illumination using the probabilities $x_n = x = 0.2$ and $y_n = y = 0.1$

The distribution function of the DR intermediates is a simple exponential function. Therefore, the average chain length of the DR intermediates is simply $\bar{n}(DR) = 1 + 1/x$. The situation of the AC and SO molecules is more complex. The average chain length is $\bar{n}(AC) = 1 + 1/x + 1/y$ and $\bar{n}(SO) = \bar{n}(AC) + 1$. The distributions shown in Fig. 23 are only a rough approximation to the experimental observations; they clearly show the rapid decrease of the absorption intensities at longer chain length observed in the experiments of the low temperature photopolymerization process. A more detailed analysis has ben given by Gross et al. [85].

# 5 Primary Processes

## 5.1 Dimer Initiation

The energy level scheme of the monomer diacetylene TS-6 crystals is shown in Fig. 24. The positions of the excited singlet (S) and triplet states (T) of the diacetylene unit $-C\equiv C-C\equiv C-$ follow from the optical absorption [86] and emission spectra [87] of the pure monomer crystals. In most cases the UV excitation of the monomer crystal is performed in the extreme absorption edge of the $S_0 \rightarrow S_1$ absorption in order to obtain a very homogeneous distribution of reaction centers M* throughout the crystal. The energy then is expected to relax very rapidly into the lowest excited triplet state $S_1 \rightarrow T_1$ via intersystem crossing and internal conversion processes. In TS-6 the lowest excited states of the substituents R lie above the corresponding excited singlet and triplet states of the diacetylene unit.

At low temperatures the polymerization process is only initiated by UV-, $\gamma$- or x-irradiation via excitation of the monomer molecules. However, in order to understand the relatively high quantum yield obtained in the bimolecular initiation reaction of Eq. (9) we have to consider a metastable long-lived excited state M*, which represents

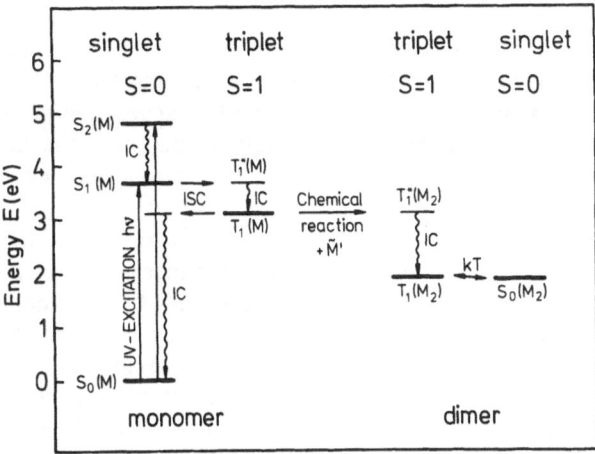

**Fig. 24.** Energy level diagram of the photophysical and photochemical primary processes of the low temperature photopolymerization reaction in diacetylene crystals. IC: Internal Conversion; ISC: Intersystem Crossing

the reaction center of the initiation reaction. The monomer diradical $T_1(M) = DR_1$ of Fig. 24 with the structure $R—\dot{C}=C=C=\dot{C}—R$ is a good approximation of the lowest excited $\pi\pi^*$ singlet and $\pi\pi^*$ triplet state of the diacetylene molecule in the *trans* configuration. The excitation energy of a monomer dicarbene $DC_1$ with the structure $R—\dot{C}—C\equiv C—\dot{C}—R$ is very high in energy (due to the additional loss of a $\pi$-bond) and is, therefore, not considered in the discussion. Even upon direct excitation into the dicarbene state the molecule is expected to relax very rapidly into the lowest excited $T_1$ state.

The most probable photochemical primary process of the polymerization reaction is shown in the energy level diagramm of Fig. 24. In the photoinitiation reaction an excited adjacent monomer molecule $M'^*$ is added to the reaction center, best represented by the metastable triplet $DR_1$ monomer molecule. Owing to the spin conservation rules, we conclude that an excited *triplet* dimer- diradical $DR_2$ is formed in the chemical reaction. We may, therefore, formulate the reaction as follows:

$$DR_1(T_1) + M'^* \rightarrow DR_2(T_1) \overset{kT}{\rightleftharpoons} DR_2(S_0) \qquad (22)$$

The triplet dimer diradical $DR_2(T_1)$ finally will relax into thermal equilibrium (kT) with its singlet ground state $DR_2(S_0)$. As we have seen from the ESR spectra (see Fig. 10) the energy separation between the singlet and triplet diradical states is very low and thermally activated transitions occur even at low temperatures. Furthermore the ESR spectra have revealed an admixture of about 10% carbene character with the diradical intermediates. This carbene character may be important in determining the probability x of the side reactions (see Eq. (19)) for the $DR \rightarrow AC$ chain termination reaction. It surely is not, however, the only essential factor, otherwise there should be no difference in the optical and thermal termination reaction steps. Up to now a direct observation of the metastable triplet state $T_1(M)$ has been possible only in two specific crystals [87, 88], where the polymerization reactions are very weak.

The dimer initiation is a multistep process with two essential steps, the creation of the reaction center "$DR_1$" with reactive electrons and the addition of an excited adjacent molecule $M'^*$. As we know from the thermal addition reaction steps at low temperatures the nature of the "second" excitation in the bimolecular reaction must be extremely low in energy and, therefore, electronic excitations may be excluded. On the other hand molecular motions are likely to be more important in the dimer formation process. Therefore we have to inspect the topochemistry of the reaction. The most probable mechanism of the dimer formation is shown in Fig. 25. In the first step (1) (only by optical excitation) the metastable reaction center is generated. In the second step (2) the motion required in the reaction is performed by librational excitation of the adjacent molecules. This librational excitation may be thermally activated and therefore is present in the crystals at elevated temperatures only. At low temperatures (T < 80 K) this librational excitation is generated only by absorption of photons with subsequent radiationless relaxation processes, producing phonons and librons. A detailed analysis of the dimer initiation process has been given by Neumann et al. [89].

The low activation energy of the thermal addition polymerization reaction confirms the necessity of a (librational) motion of the molecules in the initiation process. The first addition process differs from all the following addition proccesses by the metastable monomer diradical structure, which — in contrast to the $DR_n$, $AC_n$, and $DC_n$ structures with n > 2 — has a limited life-time given by the phosphorescence decay of the monomer triplet state. Therefore, the librational excitation must be performed during the life-time of the monomer reaction centre. In the case of the low temperature photopolymerization reaction the librational excitation has to be prepared optically via the decay of the electronic excitation. This is in contrast to the photopolymerization reaction at high temperatures, where numerous molecular motions are thermally and stationary present in the crystals. Due to this difference two photons (2hv) are required in every dimer initiation process at low temperatures and only one photon (hv + kT) is required at high temperatures. The two paths of the photoinitiation reaction are illustrated below by the arrows in Fig. 26. The respective pair states are characterized by $M^*M'^*$ and $M^*M'$ as discussed below.

In recent experiments the photopolymerization process has been initiated with visible light. Sensibilization of the photopolymerization reaction is possible in diacetylene crystals by introduction of energetically low lying absorptions of the substituents via formation of mixed crystals or by doping with dye molecules [90]. Although the detailled mechanism of the sensibilization is not clear, the experiments clearly demonstrate the importance of lowlying electronic states in the polymerization reaction.

## 5.2 Stability and Reactivity

In the discussion of the chain initiation and chain propagation reactions we need to consider the binding energies of the different DR, AC, DC intermediates and SO oligomer molecules. In first order they are obtained by simply summing up the energies of their π- and σ-bonds. From energetic considerations further informations concerning the paths of the individual polymerization reaction steps can be obtained.

In the $DR_2$ formation process two monomer molecules are connected by a single σ-bond yielding an energy of 3.6 eV. Simultaneously two monomer π-bonds with an energy of $2 \times 2.7$ eV $= 5.4$ eV are disrupted. Consequently the net energy loss in the dimer $DR_2$ formation process amounts to $E(DR_2) = 1.8$ eV.

In the $DC_2$ formation process two monomer molecules are connected by a double bond, yielding the energy of a (σ + π)-bond. Simultaneously four monomer π-bonds are lost in the reaction. Consequently, according to (3π − 1σ)-bonds, the net energy loss in the dimer $DC_2$ formation process amounts to $E(DC_2) = 4.5$ eV. Due to this high energy no dimer dicarbenes are observed!

In the $AC_2$ formation process via a $DC_2$ excited state and simultaneous intramolecular insertion reaction following Eq. (8b) first the energy of the $DC_2$ formation is required. However, in the subsequent insertion reaction the energy of a π-bond is gained. Therefore the net energy amount is $E(AC_2$-intramol. insertion) $= (DC_2) -$ $- E(\pi) = 4.5$ eV $- 2.7$ eV $= 1.8$ eV.

The $AC_1$ formation seems more probable than the $AC_2$ formation in an intramolecular insertion reaction. In this situation a $DC_1$ excited state is required, which reacts to the $AC_1$ state. The net energy amount of this reaction therefore is $E(AC_1$-intramol. insertion) $= E(DC_1) - E(\pi) = (2 \cdot 2.7 - 2.7)$ eV $= 2.7$ eV.

In the $AC_2$ formation process, by an intermolecular $CH_2$ insertion reaction (8a) a σ-bond is formed upon disruption of two π-bonds. Therefore as with the $DR_2$ molecules the net energy loss amounts to $E(AC_2$-intermol. insertion) $= 1.8$ eV. Upon formation of a pseudocyclopropene structure (7) three π-bonds are disrupted and two distorted σ-bonds (τ-bonds) with about 5 eV [91] are formed. In this case the net energy loss amounts to $E(AC_2$-pseudocyclopropen) $\cong 3$ eV.

In the bimolecular UV-photoinitialization reaction, due to the high energy of the monomer singlet excitation, an energy of $2 \times 4$ eV is available. According to the procedure described above, however, only one reaction centre is necessary, which has been interpreted as being the diradical monomer triplet state with an energy of about 3 eV. Compared to this value the librational excitation energy is negligibly small. With this energy of 3 eV only the DR- and AC-dimer but not the DC-dimer reactions are possible as observed experimentally.

In the chain propagation reaction a π-bond is changed into a σ-bond by addition of an adjacent monomer molecule to the intermediates. In this way in every reaction step about 0.9 eV are released. The resulting energy level scheme of the polymerization reaction is shown in Fig. 26. It represents the energetic positions of the resulting DR or AC intermediates characterized by the general notation $M_n$ and the transient pair states $M_nM'^*$ and $M_n\tilde{M}'$. The addition reaction steps may be induced optically (hv) or thermally (kT). In both cases, according to our reaction model of Fig. 25, the reaction is presumed to pass the librational excitation $\tilde{M}'$ of the adjacent molecules via the relaxation step $M'^* \to \tilde{M}'$. The small activation energies of the individual addition reaction steps at short chain lengths characterize the addition reaction, in which no further electronical excitation is required due to the presence of reactive radical or carbene chain ends of the DR, AC and DC intermediates.

In the thermal addition reaction a structural change from the butatriene to the acetylene chain structure is observed at n = 6 in the ESR spectra of the short intermediates. Since the ESR is sensitive only to unpaired electrons with $S \neq 0$ the butatriene-to-acetylene transformation is observed only in the thermally excited DR and DC states and not in their corresponding ground states. The low activation energies, however, suggest an almost identical chain structure in the $S = 0$, $S = 1$ or $S = 2$ states. Moreover from the optical spectra, which are sensitive to the chain structure only, it is not possible to distinguish between states of different multiplicity. By simply counting the σ- and π-bonds the DR intermediates should be more stable than the DC intermediates by an amount of 2.7 eV, corresponding to the additional π-bond. This energy difference obviously is compensated by a slightly higher stability

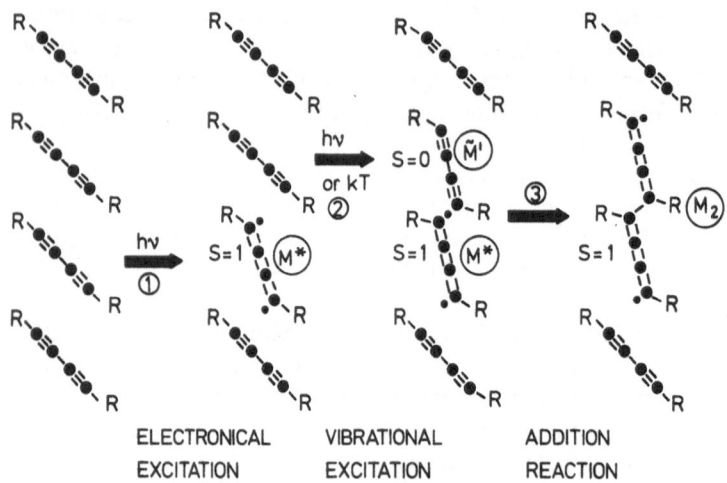

Fig. 25. Reaction scheme of the dimer initiation reaction. (1) Formation of the metastable monomer diradical M*; (2) Distortion of the adjacent monomer molecule M'; (3) Formation of the diradical dimer molecule $M_2$ by 1,4 addition

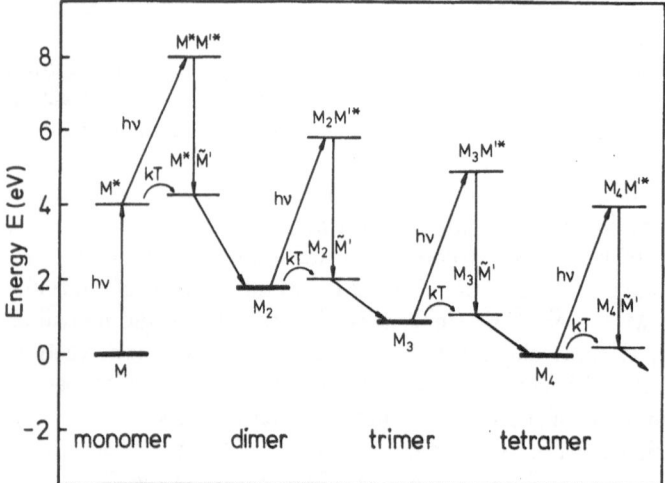

Fig. 26. Energy level scheme representing the initial steps of the polymerization reaction. The two paths of the optical (hν) and thermal (kT) addition reaction are characterized by the arrows

of the acetylene units $\pm RC-C\equiv C-CR\pm$ as compared to the butatriene units $+RC=C=C=CR+$, which have the same total number of $\sigma$- and $\pi$-bonds. An energy difference of about 0.4 eV per unit easily could explain the observed structural change at about n = 6. This butatriene-to-acetylene transformation has been observed also in the TS-12 and TCDU diacetylene crystals [92-94] at about the hexamer unit.

The binding energies, gained in the termination reaction DC → AC and AC → SO by saturation of the carbene ends, are identical. However the insertion reactions (8) with an energy gain of 3.6 eV are energetically favoured as compared to the pseudo-cyclopropene reaction with only 2.3 eV. In the DR → AC termination reaction the radical electron pair is maintained and therefore only a low change in the stability is expected. However, energy is gained by the structural butatriene-to-acetylene trans-formation. Again the insertion reactions (with n · 0.4 eV + 0.9 eV) are energetically favoured as compared to the pseudocyclopropene reaction (with n · 0.4 eV − 0.4 eV). The optically induced chain terminations are very efficient. However, thermal chain termination reactions have not been observed with the short chain intermediates with n ≲ 10, which may be resolved either in the optical or ESR absorption spectra. Consequently, in contrast to the pure photopolymerization reaction at low tem-peratures much longer polymer chains with n ≈ 150 [68] are obtained in the high temperature photopolymerization reaction, where — after photoinitiation — sub-sequent thermal addition polymerization reactions are very effective.

# 6 Conclusions

From the experimental results and conclusions drawn from the investigations of the TS-6 model system we may be able to derive general characteristics of the poly-merization reaction in diacetylene crystals. This is due to the fact, that — as a rule — the diradical, carbene and dicarbene intermediates are not directly dependent on the structure and chemical properties of the substituents. For instance, in BPG [88] and TCDU [47] crystals carbene states (located at the ends of long oligomer molecules) have been also detected in the ESR spectra, which are very similar to those in TS-6. The similarity of the optical absorption spectra of the short chain intermediates is demonstrated in Fig. 8 which shows the same diradical series A, B, C, ... in TCDU, TCDA and TS-12 [26] as observed in the TS-6 crystals even at almost the same energetic positions. Differences in the polymerization reactions in the differently substituted diacetylene crystals essentially arise from the specific molecular arrangements, resulting in specific changes of the distances and orientations of the reactive centres of the individual molecules. Therefore, the distinct reactivities of the different diacety-lene crystals are predetermined by their distinct topochemistry of essentially the dimer initiation reaction, which dominates the overall quantum yield of the polymerization reaction.

Owing to the bimolecularity of the initiation reaction the quantum yield of the dimer molecules ($M_2/N_{abs}$) is proportional to the absorbed light quanta $N_{abs}$ and to the ratio $k_1/k_0$, characterizing the competition of the chemical dimer initiation process ($k_1$) with the deactivation processes ($k_0$) of the monomer excitation. A com-parison of the dimer A absorption intensities of different diacetylene crystals shows that the ratio $k_1/k_0$ is about a factor of $10^2$ to $10^3$ larger in the TS-6 crystals than in the TS-12, TCDU or BPG crystals. The high photoreactivity of our model system TS-6 therefore, gives good experimental conditions for a detailed analysis of all the different reaction intermediates. Nevertheless, the quantum yield of about $10^{-2}$ (that is 100 photons per reaction step [56]) in TS-6 crystals is low as compared to the photochemical processes using silver halides and, therefore, at the moment is only of minor importance for technical applications.

In this article it has been shown, that the low temperature photopolymerization reaction of diacetylene crystals is a highly complex reaction with a manifold of different reaction intermediates. Moreover, the diacetylene crystals represent a class of material which play a unique role within the usual polymerization reactions conventionally performed in the fluid phase. The spectroscopic interest of this contribution has been focussed mainly on the electronic properties of the different intermediates, such as butatriene or acetylene chain structure, diradical or carbene electron spin distributions and spin multiplicities. The elementary chemical reactions within all the individual steps of the polymerization reaction have been successfully investigated by the methods of solid state spectroscopy. Moreover we have been able to analyze the physical and chemical primary and secondary processes of the photochemical and thermal polymerization reaction in diacetylene crystals. This success has been largely due to the stability of the intermediates at low temperatures and to the high informational yield of optical and ESR spectroscopy in crystalline systems.

*Acknowledgements*: The work has been supported by the Deutsche Forschungsgemeinschaft and by the Stiftung Volkswagenwerk.

Some investigations have been performed in collaboration with chemists (Prof. G. Wegner, Freiburg) and physicists (Prof. M. Schwoerer, Bayreuth and Prof. H. C. Wolf, Stuttgart). Helpful discussions with C. Bloor, D. Batchelder, H. Gross, W. Neumann, D. Siegel, and K. Ulrich are gratefully acknowledged.

# 7 References

1. Wunderlich, B.: Macromolecular Physics, Vol. 2, N.Y., Academic Press, 1976
2. Uhlmann, D. R. and Kolbeck, A. G.: Sci. Am. *233*, 96 (1975)
3. Keller, A.: Phil. Mag. *2*, 1171 (1957)
4. Geil, P. H.: Polymer Single Crystals, Interscience, New York 1963
5. Wegner, G.: Z. Naturforschung *24b*, 824 (1969)
6. Wegner, G.: Die Makromoleculare Chemie *134*, 219 (1970) and *154*, 35 (1972)
7. Wegner, G.: Chimia *28*, 475 (1974)
8. Wegner, G.: in Molecular Metals (ed. by Hatfield, W. E.), Plenum Press, N.Y. 209 (1979)
9. Wegner, G.: in Chemistry and Physics of One-Dimensional Metals (ed. by Keller, H. J.), Plenum Press, N.Y. 297 (1977)
10. Baughman, R. H.: J. Appl. Phys. *43*, 4362 (1972)
11. Baughman, R. H.: J. Polym. Sci., Polym. Phys. Ed. *12*, 1511 (1974)
12. Baughman, R. H. and Yee, K. C.: J. Polym. Sci., Polym. Chem. Ed. *12*, 2467 (1974)
13. Baugham, R. H. and Chance, R. R.: in Synthesis and Properties of Low-Dimensional Materials (ed. by Miller, J. S. and Epstein, A.), Acad. Science, N.Y. 705, (1978)
14. Bloor, D.: in Developments in Crystalline Polymers (ed. by Bassett, D. C.), Appl. Sci. Publ. London, 1982 in press
15. Wunderlich, B.: Adv. Polym. Sci. *5*, 566 (1968)
16. Geserich, H. P. and Pintschovis, L.: Festkörperprobleme *16*, 65 (1976); Street, G. B. and Greene, R. L.: IBM Res. Development *21*, 99 (1977)

17. Lochner, K., Reimer, B., and Bässler, H.: Chem. Phys. Lett. *41*, 388 (1976); Phys. Stat. Sol. (b) *76*, 533 (1976)
18. Bloor, D. and Preston, F. H.: Phys. Stat. Sol. (a) *37*, 427 (1976)
19. Reimer, B. and Bässler, H.: Chem. Phys. Lett. *43*, 81 (1976); Phys. Stat. Sol. (a) *32*, 435 (1975); (b) *85*, 145 (1978)
20. Wilson, E. G.: J. Phys. C *8*, 727 (1980); *13*, 2885 (1980)
21. Siddiqui, A. S.: J. Phys. C *13*, 2147 (1980)
22. Greene, R. L., Street, G. B., and Stuter, L. J.: Phys. Rev. Lett. *34*, 577 (1975)
23. Baughman, R. H., Chance, R. R., and Cohen, M. J.: J. Chem. Phys. *64*, 1869 (1976); Baughman, R. H., Chance, R. R.: J. Polym. Sci., Polym. Phys. Ed. *14*, 2019 (1976)
24. Leyrer, R. J., Wegner, G., and Wettling, W.: Ber. Bunsenges. Phys. Chem. *82*, 697 (1978)
25. Mondong, R. and Bässler, H.: Chem. Phys. Lett. *78*, 371 (1981)
26. Wenz, G. and Wegner, G.: Makromol. Chem. Rapid. Comm., to be published; Siegel, D., Sixl, H., Enkelmann, V., and Wenz, G.: Chem. Phys. *72*, 201 (1982)
27. Bloor, D., Koski, L., Stevens, G. C., Preston, F. H., Ando, D. J.: J. Mat. Sci. *10*, 1678 (1975)
28. Niederwald, H., Richter, K.-H., Güttler, W., and Schwoerer, M.: Laser Chemistry (1983) to be published
29. Kohlschütter, H. W.: Z. anorg. Chem. *105*, 121 (1918)
30. Thomas, J. M.: Phil. Trans. Roy. Soc. A *277*, 251 (1974)
31. Cohen, M. D.: Angew. Chem. *87*, 439 (1975)
32. Meyer, W., Lieser, G., and Wegner, G.: J. Polymer Sci. Polym. Phys. Ed. *16*, 1365 (1978)
33. Nakanishi, H., Hasegawa, M., and Sasada, Y.: J. Polymer Sci. A 1, *7*, 735 (1969), and A 2, *10* 1573 (1972)
34. Kobelt, D. and Paulus, E. F.: Acta crystallogr. B, *30*, 232 (1974)
35. Enkelmann, V. and Wegner, G.: Makromol. Chem. *178*, 635 (1977)
36. Enkelmann, V.: Acta Cryst. B *33*, 2842 (1977)
37. Enkelmann, V. and Wegner, G.: Angew. Chem. *89*, 432 (1977)
38. Enkelmann, V., Leyrer, R. J., and Wegner, G.: Makromol. Chem. *180*, 1787 (1979)
39. Bloor, D.: Europhys. News *8*, 1 (1977)
40. Bloor, D., Preston, F. H., and Ando, J.: Chem. Phys. Lett. *38*, 33 (1976)
41. Reimer, B., Bässler, H., Hesse, J., and Weiser, G.: Phys. Stat. Sol. (b) *73*, 709 (1976)
42. Bloor, D., Ando, D. J., Preston, F. H., and Batchelder, D. N.: in Structural Studies of Macromolecules by Spectroscopic Methods (ed. by K. J. Ivin), Wiley (1976) 91
43. Bloor, D. and Preston, F. H.: Phys. Stat. Sol. (a) *39*, 607 (1977); *40*, 279 (1977)
44. Hersel, W.: Thesis Univers. Stuttgart (1981)
45. Kuhn, H.: J. Chem. Phys. *17*, 1198 (1949)
46. Exarhos, G. J., Risen, jr., W. M., and Baughman, R. H.: J. Am. Chem. Soc. *98*, 481 (1976)
47. Gross, H., Sixl, H., Kröhnke, C., and Enkelmann, V.: Chem. Phys. *45*, 15 (1980)
48. Bloor, D., Ando, D. J., Preston, F. H., and Stevens, G. C.: Chem. Phys. Lett. *24*, 407 (1974)
49. Stevens, G. C. and Bloor, D.: J. Polym. Sci. Polym. Phys. Ed. *13*, 2411 (1975); Chem. Phys. Lett. *40*, 37 (1976)
50. Eichele, H., Schwoerer, M., Huber, R., and Bloor, D.: Chem. Phys. Lett. *42*, 342 (1976)
51. Niederwald, H., Eichele, H., and Schwoerer, M.: Chem. Phys. Lett. *72*, 242 (1980)
52. Niederwald, H, and Schwoerer, M.: Z. Naturforsch. *38a*, 749 (1983); Niederwald, H.: Thesis, Univers., Bayreuth (1982)
53. Hori, Y. and Kispert, L. D.: J. Chem. Phys. *69*, 3826 (1978)
54. Hori, Y. and Kispert, L. D.: J. Am. Chem. Soc. *101*, 3173 (1979)
55. Bubeck, C., Sixl, H., and Wolf, H. C.: Chem. Phys. *32*, 231 (1978)
56. Bubeck, C., Neumann, W., and Sixl, H.: Chem. Phys. *48*, 269 (1980)
57. Sixl, H., Hersel, W., and Wolf, H. C.: Chem. Phys. Lett. *53*, 39 (1978)
58. Hersel, W., Sixl, H., and Wegner, G.: ibid. *73*, 28 (1980)
59. Wegner, G.: Pure Appl. Chem. *49*, 443 (1977)
60. Wegner, G., Arndt, G., Graf, H.-J., and Steinbach, M.: in Reactivity of Solids (ed. by Wood, J., Lindqvist, O., Helgesson, C., and Vannerberg, N.-G.), Plenum Press 487 (1977)
61. Chance, R. R. and Patel, G. N.: J. Polym. Sci. Polym. Phys. Ed. *16*, 859 (1978)
62. Neumann, W. and Sixl, H.: Chem. Phys. *50*, 273 (1980)
63. Neumann, W. and Sixl, H.: ibid. *58*, 303 (1981)

64. Huber, R. and Schwoerer, M.: Chem. Phys. Lett. *72*, 10 (1980)
65. Bubeck, C., Hersel, W., Sixl, H., and Waldmann, J.: Chem. Phys. *51*, 1 (1980)
66. Huber, R. A., Schwoerer, M., Benk, H., and Sixl, H.: Chem. Phys. Lett. *78*, 416 (1981)
67. Hartl, W. and Schwoerer, M.: Chem. Phys. *69*, 443 (1982)
68. Gross, H. and Sixl, H.: Mol. Cryst. Liq. Cryst. *93*, 261 (1983); Gross, H.: Thesis, Univers. Stuttgart (1983)
69. Gross, H. and Sixl, H.: Chem. Phys. Lett. *91*, 262 (1982)
70. Brédas, J. L., Chance, R. R., Silbey, R., Nicolas, G., and Durand, Ph.: J. Chem. Phys. *75*, 255 (1981)
71. Hädicke, E., Mez, E. C., Krauch, C. H., Wegner, G., and Kaiser, J.: Angew. Chem. *83*, 253 (1971)
72. Huber, R., Schwoerer, M., Bubeck, C., and Sixl, H.: Chem. Phys. Lett. *53*, 35 (1978)
73. Bernheim, R. A., Kempf, R. J., Gramas, J. V., and Skell, P. S.: J. Chem. Phys. *43*, 196 (1965)
74. Bernheim, R. A., Bernhard, H. W., Wang, P. S., Wood, L. S., and Skell, P. S.: J. Chem. Phys. *53*, 1280 (1970)
75. Hutchison, jr., C. A. and Kohler, B. E.: J. Chem. Phys. *51*, 3327 (1969)
76. Murray, R. W. and Trozzolo, A. M.: J. Org. Chem. *29*, 1268 (1964)
77. Bubeck, C.: Thesis Univers. Stuttgart (1979)
78. Kirmse, W.: Carbene Chemistry, Academic Press, N.Y. (1971)
79. Jones, J. J. M.: Sci. Am. *2*, 101 (1976)
80. Benk, H. and Sixl, H.: Mol. Phys. *42*, 779 (1981)
81. Schwoerer, M., Huber, R. A., and Hartl, W.: Chem. Phys. *55*, 97 (1981)
82. Kollmar, C., Sixl, H., Benk, H., Denner, V., and Mahler, G.: Chem. Phys. Lett. *87*, 266 (1982)
83. Neumann, W.: Thesis, Univers. Stuttgart (1983)
84. McGhie, A. R., Kalayanaraman, P. S., and Garito, A. F.: J. Polym. Sci. Polym. Lett. Ed. *16*, 335 (1978)
85. Gross, H., Neumann, W., and Sixl, H.: Laser Chemistry (1983) in press
86. Kawaoka, K.: Chem. Phys. Lett. *37*, 561 (1976); Hardwick, J. L. and Ramsay, D. A.: Chem. Phys. Lett. *48*, 399 (1977)
87. Bertault, M., Fave, J. L., and Schott, M.: Chem. Phys. Lett. *62*, 161 (1979)
88. Bubeck, C., Sixl, H., Bloor, D., and Wegner, G.: Chem. Phys. Lett. *63*, 574 (1979)
89. Gross, H., Neumann, W., and Sixl, H.: Chem. Phys. Lett. *92*, 584 (1983)
90. Tieke, B. and Wegner, G.: Makromol. Chem. *179*, 2573 (1978); Enkelmann, V., Tieke, B., Knapp, H., Lieser, G., and Wegner, G.: Ber. Bunsenges. Phys. Chem. *82*, 876 (1978); Braunschweig. F. and Bässler, H.: ibid. *84*, 177 (1980); Bubeck, C., Tieke, B., and Wegner, G.: ibid. to be published
91. Lathan, W. A.: Structures and Stabilities of Three Membered Rings in Topics Curr. Chem., Vol. 40, 1 (1973), Springer

H.-J. Cantow (Editor)
Received July 26, 1983

# Structural Aspects of the Topochemical Polymerization of Diacetylenes

Volker Enkelmann
Institut für Makromolekulare Chemie, Hermann-Staudinger-Haus,
Stefan-Meier-Str. 31, 7800 Freiburg, FRG

*The present state of knowledge of the topochemical polymerization of diacetylenes is reviewed with regard to structural properties. Principles and mechanisms of topochemical reactions are described. Solid-state polymerization of diacetylenes utilizes the special packing properties of monomer units in their crystals. Rules relating reactivity in the solid-state and packing properties are given and the relation between reaction mechanism, molecular mobility and polymer morphology are discussed. The growth of the macromolecules in the monomer crystal, the molecular weight and its distribution with regard to the reaction kinetics and phase transitions are outlined. The electronic structure of the polydiacetylene chain is best represented by the acetylene resonance structure consisting of alternating double, single and triple bonds.*

Advances in Polymer Science 63
© Springer-Verlag Berlin Heidelberg 1984

# 1 Introduction

Many examples of solid-state reactions in organic crystals already were reported in the nineteenth century. Since then a great variety of examples have been found, e.g. dimerizations, polymerizations, cis-trans isomerizations or substituent migrations. These fascinating and, at the time of their discovery inexplicable results have been only recently understood using the rapid progress of x-ray structure analysis and spectroscopic techniques. Diacetylenes have been noted for a long time to undergo drastic colour changes upon storage or exposure to light although both crystal shape and chemical analysis apparently did not change [1-11]. G. Wegner demonstrated in 1969 that the solid-state polymerization of diacetylenes can be characterized as a diffusionless, totally lattice controlled process according to [12].

$$\tag{1}$$

The principles of such topochemical reactions have been established by G. M. J. Schmidt for the dimerization of cinnamic acid derivatives which represents the other classical organic solid-state reaction [13, 14].

Some important aspects of topochemical polymerizations can be understood by inspection of Eq. (1). All reactivity comes about by very specific rotations of the monomers and by 1,4-addition of adjacent units and an extended, fully conjugated polymer chain is formed. The unique feature of the topochemical polymerization of diacetylenes is the fact that in many cases the reaction can be carried out as a single phase process. This leads to macroscopic, defect-free polymer single crystals which cannot be obtained, in principle, by crystallization of ready-made polymers by conventional methods. Thus, polydiacetylenes are ideal models for the investigation of the behaviour of macromolecules in their perfect three dimensional crystal lattice.

Owing to the mechanism of the topochemical reaction the polyconjugated polymer chain is of exceptional purity and stereochemical regularity. Polydiacetylene crystals are thus ideally suited to study the inherent optical and electrical properties of poly-conjugated chains. These unique features have attracted considerable attention and in recent years the topochemical polymerization of diacetylenes has developed to one of the best investigated chemical reactions. It is therefore almost impossible to cover all developments ranging from synthetic chemistry to solid state physics.

The following is an attempt to summarize the structural data which have been accumulated recently and to critically review the present state of knowledge on the formation and structural properties of polydiacetylenes and to point out some directions of future developments in the field. Other aspects have been reviewed recently [15-25] and the photopolymerization and reaction kinetics as well as the spectroscopic identification of the reactive intermediates will be covered by H. Bässler and H. Sixl [107, 116].

## 2 Principles of Topochemical Reactions

Reactivity in the solid-state is always connected with specific motions which allow the necessary contact between the reacting groups. In most cases "solid-state" reactions proceed by diffusion of reactions to centers of reactivity or by nucleation of the product phase at certain centers of disorder. This leads to the total destruction of the parent lattice. If the product is able to crystallize it is highly probable that nucleation of the crystalline product phase at the surface of the parent lattice will lead to oriented growth under the influence of surface tension. In such "topotactic" reactions certain crystallographic directions of parent and daughter phases will coincide. Typical examples for this behaviour are the solid-state polymerizations of oxacyclic compounds such as trioxane, tetroxane or β-propiolactone [26].

In contrast to this behaviour a topochemical reaction can be described as a diffusionless transformation of the parent crystal into the daughter crystal. All reactivity comes about by very specific rotations of the monomers on their lattice sites. Both crystallographic position and symmetry of the monomer units are retained in this process which is schematically shown in Fig. 1.

The principles of topochemical reactions have been worked out by G. M. J. Schmidt for the example of the well known 2+2 cycloadditions of olefins, e.g. cinnamic acid derivatives [13].

As it was mentioned above the centers of the reacting molecules ideally remain fixed in topochemical reactions. As a consequence the reaction mode which requires the smallest atom displacements will prevail (least motion principle). Reactivity is

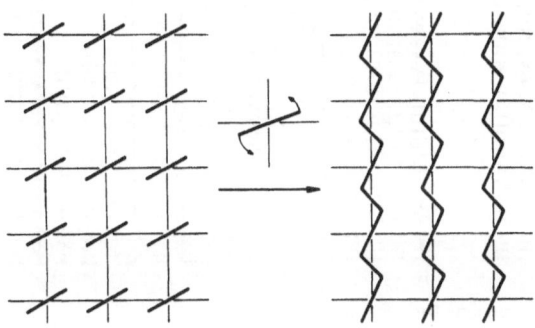

Fig. 1. Scheme of a topochemical polymerization. Transformation of a monomer single crystal into the polymer single crystal

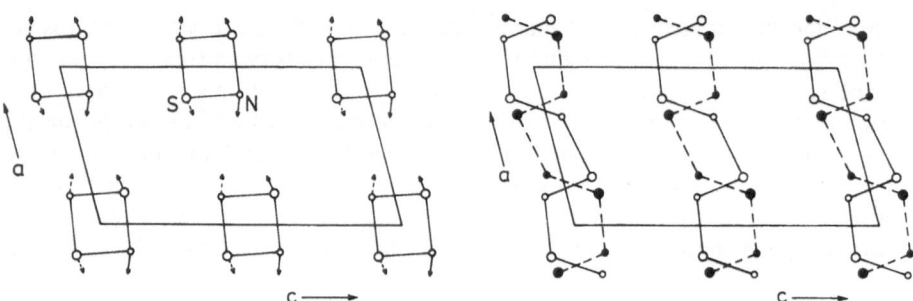

**Fig. 2.** Crystal structures of $S_2N_2$ and $(SN)_x$. The arrows in the monomer structure indicate atom displacements

only observed if the separation of the reaction atoms is less than a limiting distance of approximately 4 Å [22-25]. The new chemical bonds are all oriented along a specific crystallographic direction. For the case of a topochemical polymerization this means that extended chain polymers are formed.

The symmetry of the monomer packing determines the symmetry of the products, e.g. cinnamic acids will form only one of the various possible dimers, the symmetry of which is present in the monomer crystal [13]. Even absolute asymmetric syntheses are possible utilizing the symmetry relationships of topochemical reactions [27-29].

Topochemical reactions can be regarded as special phase transitions. If the crystal symmetry is changed during the reaction this leads, depending on the cooperativity of the molecular motion, to order-disorder structures, twinning or phase separation. A good example where this uniqueness criterion is violated is the synthesis of $(SN)_x$. The rings of the precursor $S_2N_2$ are arranged in a way that two reaction modes are equally possible (Fig. 2). As a consequence, the polymer obtained consists of fibrils with typical diameters of 200 Å [30].

Apart from the uniqueness of the reaction the morphology of the product depends substantially on the distribution of the reaction centers in the crystal. Two limiting cases can be considered. In the case of a heterogeneous topochemical process the reaction starts preferentially at specific defect sites and proceeds with nucleation of product phases. This mechanism eventually leads to the destruction of the mother crystal since the coherence between the various nuclei is lost under the influence of

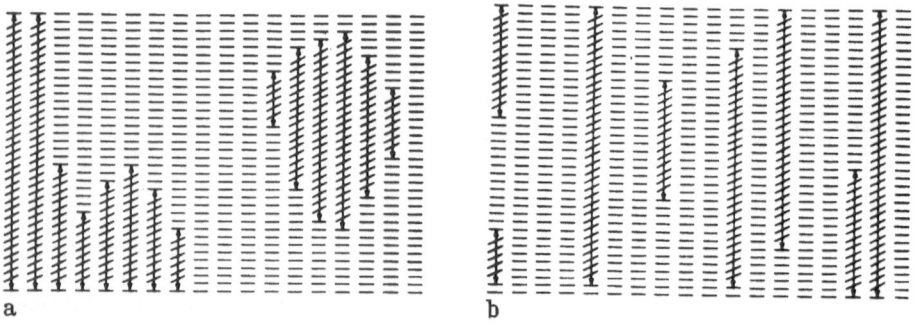

**Fig. 3a and b.** Heterogeneous (**a**) and homogeneous topochemical (**b**) reactions

the anisotropic lattice changes during the reaction [15-25, 31, 32]. Most 2+2 cyclo-additions of olefins belong to this type of topochemical reactions (Fig. 3).

In the homogeneous reaction the product is randomly distributed in the parent crystal forming a solid solution. This mechanism leads in ideal cases to single crystal transformation although in some cases large changes of lattice parameters may be observed.

Homogeneous topochemical reactions are quite rare. The topochemical polymerization of diacetylenes and with special precautions some four center photodimerizations [33, 34] are examples for this reaction mode.

# 3 Structure and Reactivity

## 3.1 Packing of Diacetylene Monomers

The topochemical polymerization of diacetylenes proceeds by a 1,4-addition reaction according to:

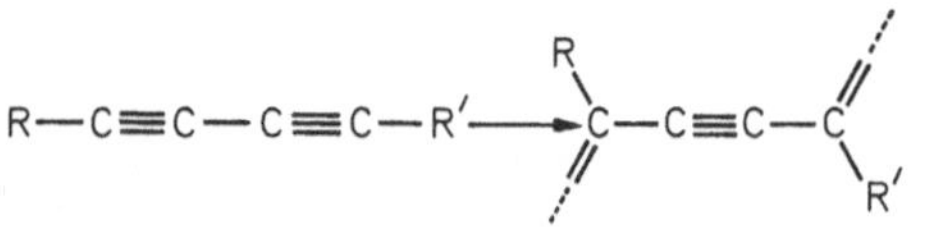

$$R-C\equiv C-C\equiv C-R' \longrightarrow \overset{R}{\underset{R'}{C}}-C\equiv C-C \tag{2}$$

The reaction is initiated by irradiation or by thermal annealing. The monomer units are arranged in a stack so that one monomer unit can react with its two neighbors. The polymer backbone which is formed in this process is oriented along a well defined lattice direction. A model of the monomer packing is schematically shown in Fig. 4. It can be characterized by the stacking distance d of the monomers in the array and by the angle $\Phi$ between the diacetylene rod and the stacking axis. In this model the

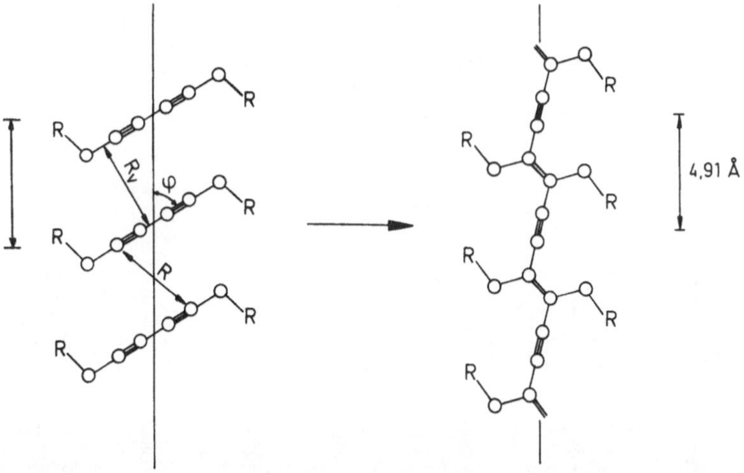

Fig. 4. Model for the packing of diacetylene monomers

**Table 1.** Packing parameters of diacetylene monomers $R-C\equiv C-C\equiv C-R'$

| R | R' | Abbr. | Space group | Point group | d/Å | Φ/° | T/K* | Reactivity Therm. | Gamma | Remark | Ref. |
|---|---|---|---|---|---|---|---|---|---|---|---|
| $-CH_2-OSO_2$—⟨benzene⟩—$CH_3$ | = R | PTS | $P2_1/c$ | $\bar{1}$ | 5.11 | 44 | 110 | +++ | +++ | A | 34–36) |
| $-CH_2-OSO_2$—⟨benzene⟩—$Cl$ | = R | PCS | $P\bar{1}$ | $\bar{1}$ | 5.03 | 67 | 295 | — | — | | 37) |
| $-CH_2-OSO_2$—⟨naphthalene⟩ | = R | PFBS | $P2_1/c$ | $\bar{1}$ | 5.07 | 44.8 | 110 | ++ | ++ | | 38) |
| $-CH_2-OSO_2$—⟨benzene⟩—$OCH_3$ | = R | PBS | $P\bar{1}$ | $\bar{1}$ | 5.03 | 65.6 | 295 | — | — | | 65) |
| $-CH_2-OSO_2$—⟨benzene⟩—$F$ | = R | MBS | $P\bar{1}$ | $\bar{1}$ | 5.80 | 62.7 | 295 | — | — | | 39) |
| $-CH_2-OSO_2$—⟨benzene⟩—$Br$ | = R | NS | $P2_1/n$ | $\bar{1}$ | 5.42 | 61.3 | 295 | — | — | | 40) |
| $-(CH_2)_2-OSO_2$—⟨benzene⟩—$CH_3$ | = R | | $P\bar{1}$ | $\bar{1}$ | 9.69 | 66.7 | 295 | — | — | | 41) |

| Compound | R | Space group | Z | | | T | | | | Ref. |
|---|---|---|---|---|---|---|---|---|---|---|
| PTS-12 | = R | 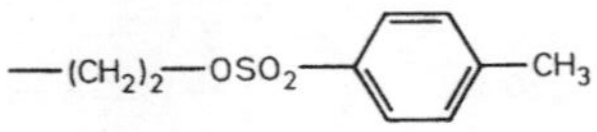 $P\bar{1}$ | 1 | 5.19 | 47.9 | 110 | — | — | B | 42) |
| BPG-1 | cycl. | 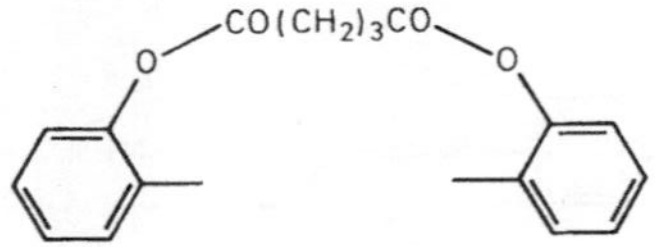 $C2/c$ | 2 | 4.93 | 46 | 110 | + | + | C | 43–45) |
| BPG-2 | | $P2_1/n$ | 1 | 5.42 | 90 | 295 | — | — | | 46) |
| DCH | = R | $P2_12_12$ | 2 | 4.66 | 90 | 295 | — | — | B | 47) |
| | = R | $P2_1/c$ | $\bar{1}$ | 4.55 | 60 | 295 | ++ | ++ | | |
| DAH | = R | $P2_1/c$ | $\bar{1}$ | 4.35 | 61.5 | 295 | — | — | D | 48) |
| ACH | —CH₂A | $P2_1/c$ | $\bar{1}$ | 4.36 | 60.8 | 295 | — | — | | 48) |
| HDU-2 | = R | 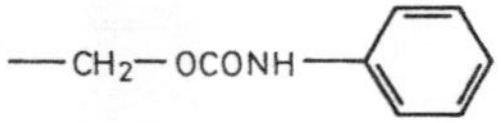 $C2/c$ | 1 | 4.00 | 66.7 | 110 | (+) | (+) | E | 31,49,50) |
| | | 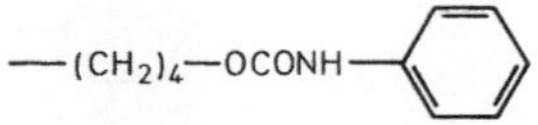 $P2_1/a$ | $\bar{1}$ | 5.18 | 42.3 | 110 | — | ++ | | |
| TCDU-1 | = R | $P2_1/c$ | $\bar{1}$ | 5.23 | 42.2 | 110 | — | + | F | 51) |
| TCDU-2 | = R | $C2/c$ | $\bar{1}$ | 5.19 | 58.3 | 110 | — | + | | 50) |
| | = R | | $\bar{1}$ | 4.63 | | 295 | (+) | (+) | E | 52) |
| | = R | $P2_1/c$ | $\bar{1}$ | 4.76 | 41 | 295 | (+) | (+) | | 53) |
| | = R | $P2_1/c$ | 1 | 4.85 | 45.5 | 120 | — | + | | 54) |
| | —CH₃ | $P2_1/c$ | $\bar{1}$ | 4.87 | 49.6 | 110 | — | + | G | 55) |
| | | 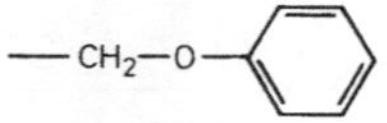 Pcab | $\bar{1}$ | 8.53 | 77 | 295 | — | — | | |
| | = R | $P2_1/c$ | $\bar{1}$ | 4.35 | 73 | 295 | — | — | | 56) |

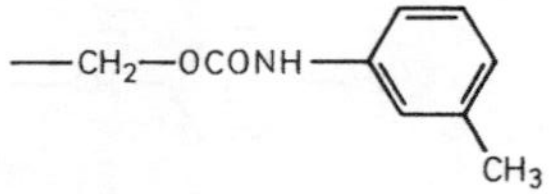

Structural formulas (R groups):

—(CH₂)₂—OSO₂—⟨C₆H₄—CH₃⟩ = R

cycl. —O—CO(CH₂)₃CO—O— (bis-o-tolyl ester)

—Cz

—CH₂—Cz

—CH₂—A

—CH₂—Cz

—CH₂—OCONH—⟨C₆H₅⟩

—(CH₂)₄—OCONH—⟨C₆H₅⟩

—CH₂—OCONH—⟨C₆H₄—CH₃⟩

—CH₂OH

—CH₂—CH₂OH

—CH₂OH  —CH₃

—CH₂—O—⟨C₆H₅⟩

**Table 1** (continued)

| R | R' | Abbr. | Space group | Point group | d/Å | Φ/° | T/K* | Reactivity Therm. | Reactivity Gamma | Remark | Ref. |
|---|---|---|---|---|---|---|---|---|---|---|---|
| $-CH_2-OCO-C_6H_5$ | =R | | $P2_1/c$ | $\bar{1}$ | 4.35 | 58 | 295 | (+) | — | H | 57) |
| | =R | | $P2_1/c$ | $\bar{1}$ | 8.48 | 40 | 295 | — | — | | |
| $-CH_2-$ (dinitrophenyl, $NO_2$, $NO_2$) | =R | DNP | $P2_1/n$ | $\bar{1}$ | 5.19 | 45.7 | 295 | ++ | + | | 58) |
| (phenyl) | =R | | $P2_1/c$ | $\bar{1}$ | 6.04 | 51 | 295 | — | — | | 59) |
| (phenyl)$-NO_2$ | $-C\equiv C-CH_3$ | | $I4_1/amd$ | mm | 7.09 | 90 | 295 | — | — | | 60) |
| $-CH_3$ | =R | | $R3m$ | 3/m | 3.8 | 90 | 295 | — | — | | 61) |
| $-SnPh_3 \times CHCl_3$ | =R | | $Pa3$ | 3 | 15.55 | 54.7 | 238 | — | — | | 62) |
| $-PbPh_3 \times CHCl_3$ | =R | | $Pa3$ | 3 | 15.40 | 54.7 | 238 | — | — | | |
| $-CH_2-N(COCH_3)-C_6H_4-N=N-C_6H_5$ | =R | | $P2_1/c$ | $\bar{1}$ | 10.79 | 23.4 | 295 | — | — | | 63) |
| $-(CH_2)_4-$ pyridinium ($N^{\oplus}$), $TCNQ^{\cdot\ominus}$ | =R | | $P1$ | $\bar{1}$ | 6.94 | 47.8 | 295 | — | — | | 64) |

| Structure | | Space group | | | | | | | | Ref. |
|---|---|---|---|---|---|---|---|---|---|---|
| —(CH₂)₃—OCONH—CH₂—COO—(CH₂)₃—CH₃ = R | 3BCMU | C2/c | 1̄ | 4.90 | 47.3 | 110 | — | ++ | F | 66) |
| CH₂—OCONH— (with Cl) = R | | P1̄ | 1̄ | 4.79 | 58.5 | 163 | — | — | H | 68) |
| —C≡C—Si(CH₃)₃ = R | | Pbcn | 1 | — | — | 295 | — | — | I | 69) |
| —(CH₂)—OCONH—CH₂—COO—(CH₂)₃—CH₃ = R | | P1̄ | 1̄ | 4.94 | 83.5 | 295 | — | — | J | 67) |
| | | | | 4.03 | 65 | | | | | |
| | | | | 4.45 | 59.5 | | | | | |
| —COOH × H₂O = R | | I2/c | 2 | 5.58 | 45 | 295 | + | + | | 4) |

\* Temperature at which the data collection for the structure analysis was carried out.

A   two independent molecules in the unit cell;
B   phase transition during the polymerization;
C   incomplete reaction;
D   order-disorder structure;
E   phase separation at a conversion of approximately 10 percent;
F   limiting conversion about 60 percent;
G   phase separation, residual monomer can be removed by sublimation;
H   low reactivity under pressure;
I   no stack structure;
J   three different intermolecular contacts in the unit cell;

—Cz: carbazolylgroup;
—A: anthrylgroup

**Table 2.** Lattice parameters of diacetylene monomers R−C≡C−C≡C−R'. E.s.d.'s are given in parentheses

| R | R' | Space group | a/Å | b/Å | c/Å | α/° | β/° | γ/° | Therm. | Gamma | Ref. |
|---|---|---|---|---|---|---|---|---|---|---|---|
| —CH₂—Cz | —CH₂OH | P2₁/c | 18.22(4) | 4.59(1) | 16.99(4) | | 99.0(3) | | − | − | 70) |
| —CH₂—N(Ph)₂ | = R | P1̄ | 8.86(1) | 10.08(1) | 14.72(1) | 110.7(1) | 111.2(1) | 81.7(1) | +++ | +++ | 71) |
| —CH₂—N(Ph)₂ | —CH₂—Cz | P2₁ | 9.1 | 29.5 | 8.8 | | 107.8 | | − | − | 70) |
| —CH₂—CH₂ TCNQ⁺⊖ (quinoline) | = R | C222₁ | 6.58(2) | 19.50(3) | 31.10(4) | | | | − | − | 70) |
| —CH₂—CH₂—N⁺ (TCNQ)⁺⊖₂ (pyridine) | = R | P1̄ | 7.76(2) | 13.87(2) | 11.33(2) | 93.1(3) | 106.8(3) | 102.4(3) | − | − | 70) |
| (benzene) NH—CO—CH₃ | = R | P1̄ | 4.71(1) | 16.10(3) | 5.24(1) | 93.7(5) | 95.0(5) | 77.0(5) | − | + | 70) |
| —(CH₂)₄—OCONH (thiophene) | = R | P2₁/c | 17.52(4) | 6.52(3) | 9.86(2) | | 97.0(3) | | − | − | 72) |
| —(CH₂)₄—OCONH—CH—CH₃ (phenyl) | = R | P1 | 9.83(3) | 13.00(3) | 12.43(3) | 92.3(3) | 91.6(3) | 73.0(3) | − | (+) | 72) |
| —(CH₂)₄—OCONH (cyclopropyl) | = R | P2₁/c | 16.87(2) | 9.14(2) | 9.55(2) | | 90.5(3) | | − | − | 72) |
| —(CH₂)₄—Cz | = R | C2/c | 19.28(1) | 13.14(1) | 10.92(1) | | 96.2(3) | | − | (+) | 71) |

| $R_1$ | $R_2$ | Space group | $a$ | $b$ | $c$ | $\alpha$ | $\beta$ | $\gamma$ | react. | react. | Ref. |
|---|---|---|---|---|---|---|---|---|---|---|---|
| $-(CH_2)_4-OSO_2-C_6H_4-OCH_3$ = R | | $P\bar{1}$ | 5.24(2) | 5.92(1) | 21.29(2) | 85.4(3) | 96.6(3) | 91.1(3) | − | (+) | 73) |
| $-(CH_2)_4-OSO_2-C_6H_4-Cl$ = R | | $P2_1/c$ | 21.19(2) | 5.18(1) | 11.53(3) | | 92.5(3) | | − | (+) | 73) |
| $-(CH_2)_9-CH_3$ | $-(CH_2)_8COOH$ | $P\bar{1}$ | 4.76(4) | 5.42(4) | 42.3(8) | 90 | 90 | 95.2(2) | − | + | 74, 75) |
| $-(CH_2)_{11}-CH_3$ | $-COOH$ | $P\bar{1}$ | 4.91(4) | 5.36(5) | 32.7(1) | 90.5(2) | 91.6(10) | 106.7(10) | − | + | 74, 75) |
| $-(CH_2)_{11}-CH_3$ | $-COOH$ ×Phenazine | $P\bar{1}$ | 4.77(4) | 7.09(5) | 33.2(1) | 91.5(2) | 91.6(10) | 104.5(10) | − | + | 74, 75) |
| $-(CH_2)_{13}-CH_3$ | $-COOH$ ×Phenazine | $P\bar{1}$ | 4.74(4) | 7.15(5) | 36.4(1) | 84.1(2) | 92.1(10) | 106.1(10) | − | + | 74, 75) |
| = R | | $P2_1/a$ | 16.82(10) | 17.72(10) | 4.70(1) | | | 96.6(6) | + | + | 73 |
| $-CH_2OCONHCH_2CH_3$ | = R | $P2_1/a$ | 9.39(4) | 17.84(10) | 8.64(3) | | | 105(2) | − | − | 76 |
| $-CH_2OH$ | $-C\equiv C-CH_2OH$ | $P2_1/c$ | 4.11 | 19.59 | 4.80 | | 109 | | (+) | ++ | 77 |
| $-C\equiv C-CH_2OCONH-Ph$ | $-CH_2OCONH-Ph$ | $P2_1/c$ | 24.65 | 30.74 | 4.89 | | 92.2 | | + | + | 77 |

approach of neighboring units is restricted by the van der Waals distance $R_v$. The polymer chain repeat distance of 4.91 Å as well as the bond lengths and angles of the polymer backbone are nearly constant for all cases studied to date (cf. Table 4). It should be emphasized at this point that there is normally a mismatch between the monomer stacking d and the polymer repeat. The consequences of the mismatch on the reaction kinetics will be discussed later.

In order to assess the relative reactivity of different diacetylene monomers the packing parameters can be analysed in terms of a reaction along a least motion reaction path, i.e. the molecules simultaneously rotate and translate along the stacking direction [22, 23]. In this model maximal reactivity is expected for d $\simeq$ 5 Å and $\Phi \simeq 45°$. In Fig. 5 the values of the packing parameters d and $\Phi$ are plotted for constant separations R between the reacting atoms C1 and C4'. The relevance of the model considerations can be tested using crystal structure data, which have become available recently for a number of reactive and unreactive diacetylene monomers. Reactivity is only observed in a small area of the map. The distribution of the points for highly reactive structures suggest the criterion for which the separation R should be less than 4 Å to be a more critical condition than the requirement of a least motion pathway as calculated by Baughman [17, 22, 23]. Figure 5 shows that all but one reactive diacetylene structure fulfill the 4 Å criterion. The exception DCH (cf. Table 1) is a special case where the reaction is connected with a phase transition. This will be discussed in detail later.

The dependence of the solid-state reactivity of diacetylenes on the packing parameters d and $\Phi$ are summarized in Table 1.

Crystallographic data of some monomers where the full crystal structure analysis has not been carried out are given in Table 2.

The experimental data qualitatively confirm the geometrical model presented in Fig. 4. The reactivity is controlled by the monomer packing and not by the chemical nature of the substituents. In many cases different modifications of a specific monomer can be obtained which exhibit drastic differences in reactivity [31]. However, it should be emphasized at this point that the packing parameters give no absolute scale for

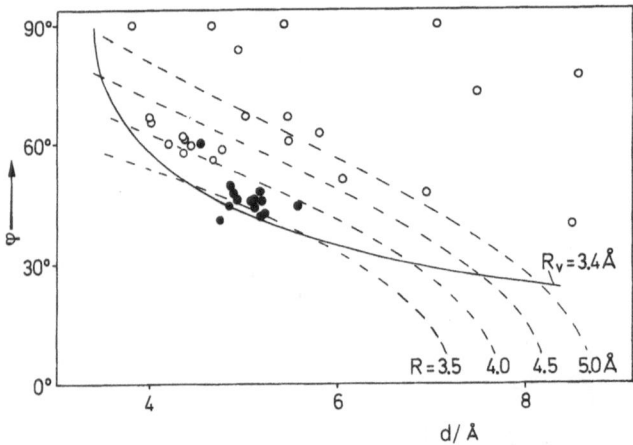

**Fig. 5.** Plot of d vs. $\Phi$. The broken lines are lines of constant distance R between C1 and C4'. ○: inactive structures, ●: reactive structures

the reaction rate. Monomers with virtually identical packing can show large reactivity differences. Some cases will be discussed in detail below.

## 3.2 Crystal Structure Analysis of Diacetylene Monomers

Until 1977 no crystal structure analyses were known for highly reactive diacetylene monomers. Polymerization in the primary x-ray beam proceeds in these cases so rapidly that data collection on the monomer crystal is impossible. This experimental difficulty was overcome by carrying out the data collection at low temperatures. At 110 K the polymerization rate is sufficiently low to maintain a polymer content at below 5 percent during the time necessary to collect the data for an average structure.

The first monomer crystal structure which was solved using this technique was PTS

$(R = R' = -CH_2OSO_2-$ ⬡ $-CH_3$ , cf. Table 1) which is regarded in many respects as

a model for the whole class of diacetylenes [34, 35]. A projection of the monomer and polymer crystal structures on a common plane is shown in Fig. 6 [78, 79].

Stereoscopic views of the packing in both structures are shown in Fig. 7.

The experimental result is in agreement with the simple geometrical model presented in Fig. 4. The packing parameters are in the range where high reactivity is expected. In a first approximation the side group packing can be considered to remain unchanged and all molecular motions are restricted to the center of the molecule. This is necessary for the formation of a solid solution of growing polymer chains in the monomer matrix. The crystallographic data given in Table 3 show that monomer and polymer crystal structures are indeed isomorphous although considerable lattice changes are involved in the reaction. These will be analysed in detail below.

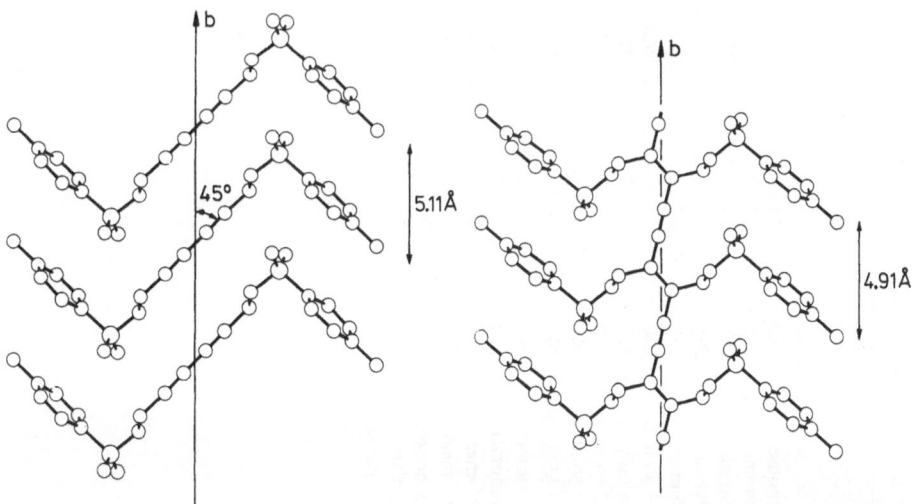

**Fig. 6.** Projection of the PTS monomer and polymer crystal structures on the plane of the polymer backbone

**Table 3.** Lattice parameters of crystal structures as discussed in detail in the text. E.s.d.'s are given in parentheses

| Structure | T/K | a/Å | b/Å | c/Å | α/° | β/° | γ/° | $D_x$ g/cm³ | Space group | Ref. |
|---|---|---|---|---|---|---|---|---|---|---|
| PTS monomer | 120 | 14.61(1) | 5.11(1) | 25.56(2) | | 92.0(5) | | 1.46 | P2₁/c | 34, 35) |
| PTS monomer | 295 | 14.60 | 5.15 | 15.02 | | 118.4 | | 1.40 | P2₁/c | 80) |
| PTS monomer | 295 | 14.65(1) | 5.178(2) | 14.94(1) | | 118.81(3) | | 1.40 | P2₁/c | 81) |
| PTS polymer | 295 | 14.993(8) | 4.910(3) | 14.936(10) | | 118.14(4) | | 1.483 | P2₁/c | 78) |
| PTS polymer | 120 | 14.77(1) | 4.91(1) | 25.34(2) | | 92.0(5) | | 1.51 | P2₁/c | 79) |
| TCDU-1 monomer | 120 | 7.08(2) | 33.97(3) | 5.23(2) | | 115.8(5) | | 1.26 | P2₁/a | 51) |
| TCDU-1 polymer | 295 | 6.229(5) | 39.027(1) | 4.909(4) | | 106.85(5) | | 1.25 | P2₁/a | 83) |
| TCDU-2 monomer | 120 | 18.97(4) | 5.19(2) | 11.60(2) | | 92.0(5) | | 1.26 | P2₁/c | 50) |
| TCDU-2 50 Mrad | 295 | 19.65(4) | 4.96(2) | 12.12(3) | | 92.5(5) | | 1.23 | P2₁/c | 50) |
| TCDU-2 65% pol. | 295 | 19.63(1) | 4.95(1) | 11.84(1) | | 94.9(1) | | 1.25 | P2₁/c | 50, 82) |
| DCH monomer | 295 | 13.60(5) | 4.55(2) | 17.60(5) | | 94.0(5) | | 1.25 | P2₁/c | 47) |
| DCH monomer | 110 | 13.38(1) | 4.20(4) | 18.44(1) | | 92.0(5) | | 1.31 | P2₁/c | 48) |
| DCH polymer | 295 | 12.87(1) | 4.91(1) | 17.40(1) | | 108.0(2) | | 1.30 | P2₁/c | 84) |
| DCH polymer | 295 | 12.821(4) | 4.886(3) | 17.328(3) | | 108.32(3) | | 1.31 | P2₁/c | 70, 85) |
| PTS-12 monomer | 110 | 20.60(2) | 11.79(1) | 5.19(1) | 83.0(3) | 89.2(3) | 92.7(3) | 1.36 | P1̄ | 42) |
| PTS-12 polymer | 110 | 20.01(2) | 6.02(1) | 4.91(1) | 95.1(3) | 93.7(3) | 88.7(3) | 1.42 | P1̄ | 42, 85) |
| PTS-12 polymer | 295 | 20.13(2) | 6.11(1) | 4.91(1) | 95.1(3) | 93.7(3) | 88.7(3) | 1.39 | P1̄ | 42) |

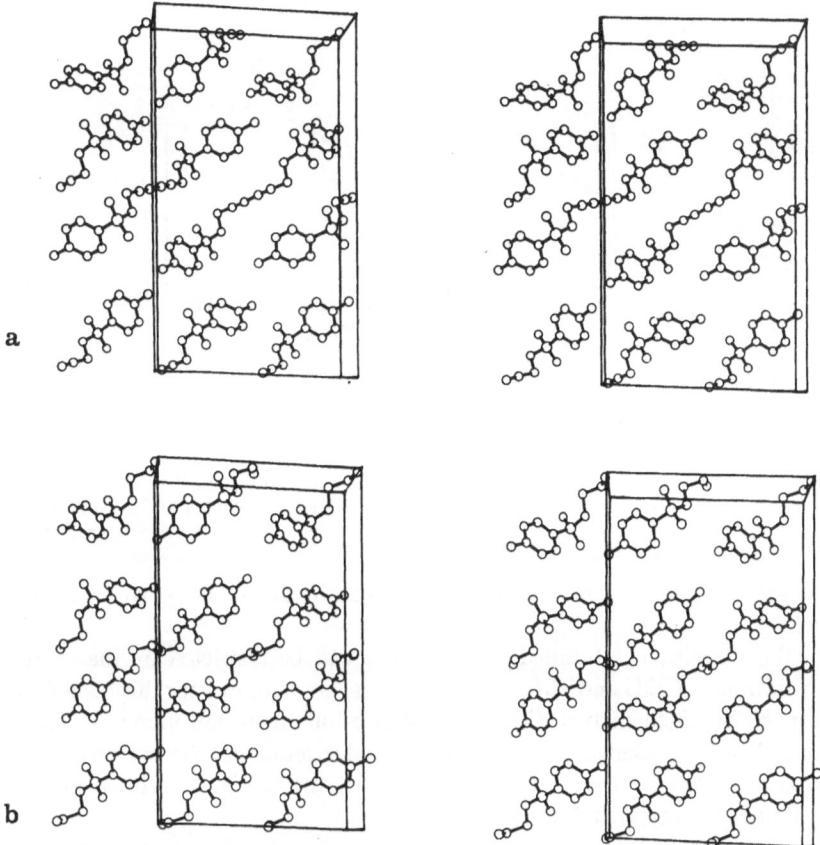

**Fig. 7a and b.** Stereoscopic projections of the packing of PTS monomer (above) and polymer (below) at 120 K. **a** is horizontal and **b** vertical

## 3.3 Molecular Motions Connected with the Topochemical Polymerization of Diacetylenes

The simple geometrical model of the least motion principle neglects the contributions of the side group packing, which is thought to remain constant during the reaction. PTS is in many respects a model compound where this ideal condition is nearly satisfied. In many other cases, however, side group packing and mobility play a dominant role in polymerization. The cyclic monomer BPG is an example where the reactivity is restricted by the side group packing. Despite of favourable monomer packing (d = 4.93 Å, Φ = 46°, Fig. 8) BPG cannot be polymerized quantitatively. At a gamma-ray dosage of 60 Mrad a limiting conversion of approximately 35 percent is reached [43, 44]. This unexpectedly, low reactivity can be qualitatively understood by the fact that the large cyclic side group is directly attached to the reactive triple bond system. This restricts the side group mobility which is necessary in addition to favourable monomer packing. Tables 1 and 2 show that in order to attain high reactivity the diacetylene group and the side group must be separated by at least a

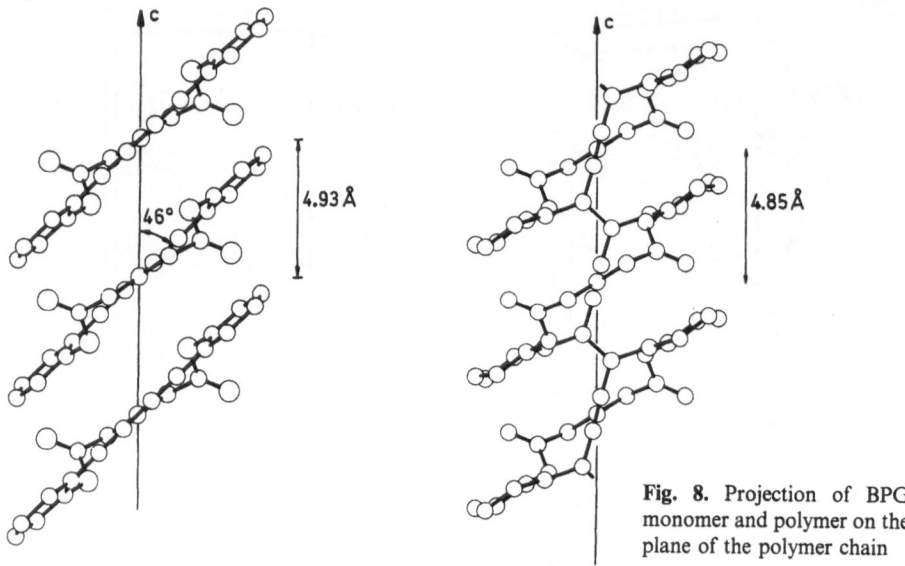

**Fig. 8.** Projection of BPG monomer and polymer on the plane of the polymer chain

methylene group acting as a mobile spacer. This empirical rule holds with very few exceptions.

Even if this condition is fulfilled the reactivity can be restricted by the mutual sidegroup motions. TCDU and 3BCMU (cf. Table 1) are examples for this behaviour. TCDU can be obtained in two different modifications in which monomer stacks with almost identical packing parameters are packed together in different ways [50, 51]. Stereoscopic projections of the crystal structures are shown in Fig. 9, pertinent crystallographic data are summarized in Table 3.

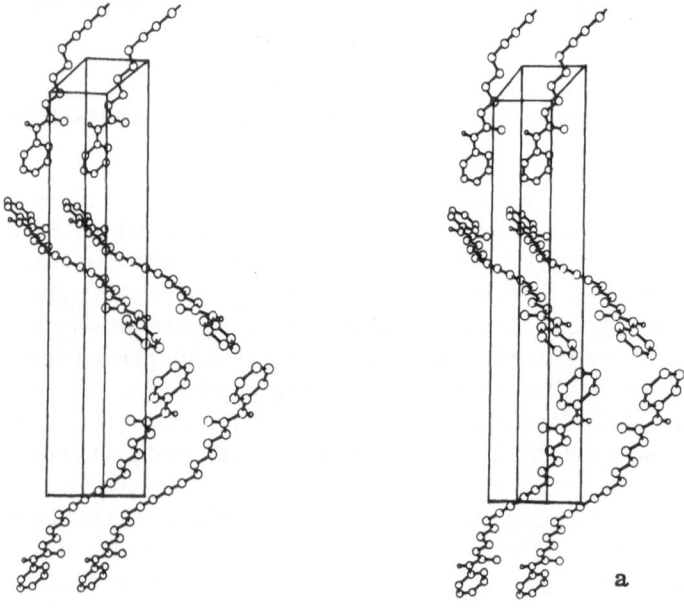

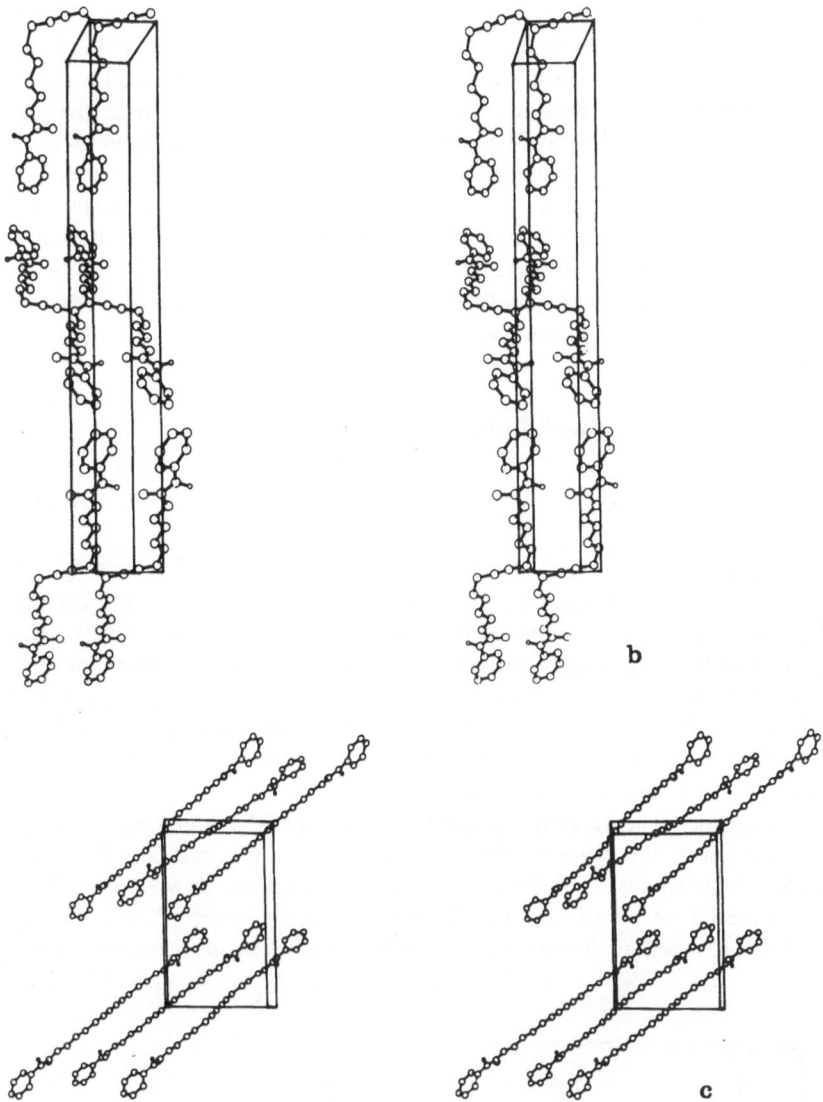

**Fig. 9a–c.** Stereoscopic projections of the packing in TCDU structures. **a)** TCDU-1 monomer, **b)** TCDU-1 polymer, **b** is vertical and **c** horizontal. **c)** TCDU-2 monomer, **a** is vertical and **c** horizontal

TCDU-1 can be polymerized with $^{60}$Co γ radiation to an almost complete conversion. However, the reaction is accompanied by unusually large rotations of the entire side chain which is evidenced by the large change of the b axis (13%, Table 3) in going from monomer to polymer. This introduces some disorder which can be detected in unusual bond lengths in the polymer structure and large thermal parameters [83].

Although TCDU-1 and TCDU-2 have very similar packing parameters TCDU-2 is much less reactive, e.g. at a dosage of 50 Mrad the stacking axis b is decreased to 4.95 Å. From this value a conversion of about 60 percent can be calculated using the known conversion dependence of the stacking distance in other reactive diacetylenes.

At this point the crystals already have suffered from radiation damage or other disorder introduced by the reaction. Higher order reflections, especially those with l > 3 begin to vanish indicating some disorder along the c-axis which roughly corresponds to the orientation of the side chain. Attempts to solve the crystal structure of the polymer with the limited available data have failed [82]. The differences in the reactivity of both modifications can be qualitatively explained by the mutual side group motions. In TCDU-2 neighboring monomer arrays are related along the c-axis by a glide plane. As a consequence of this "antiparallel" orientation, alternating sheets of molecules perform movements in opposite directions which may be the cause for the limited reactivity and the disorder introduced during the reaction. Similar observations have been made with 3BCMU [66].

## 4 Polymer Growth in the Monomer Matrix

### 4.1 Structural Investigations

PTS can be regarded in many respects as a model for the whole class of diacetylenes. The monomer can be easily prepared and obtained as large and perfect single crystals. Thermal polymerization occurs even at room temperature so that the reaction can be carried out under very mild conditions [80, 86]. Time conversion curves for three different polymerization temperatures are shown in Fig. 10. The polymerization is characterized by a slow induction period followed by a rapid reaction to complete conversion. Both reaction regimes show apparent first order kinetics with identical thermal activation energies of 92 kJ/mol [87-90].

The conversion dependence of the lattice parameters of PTS is plotted in Fig. 11. Over the entire conversion range the lattice parameters change gradually from their initial to their final values [35]. Although polymerization proceeds with quite large changes of some lattice parameters no phase separation is observed and the high crystal quality is retained in all conversions. A wide range of experimental techniques

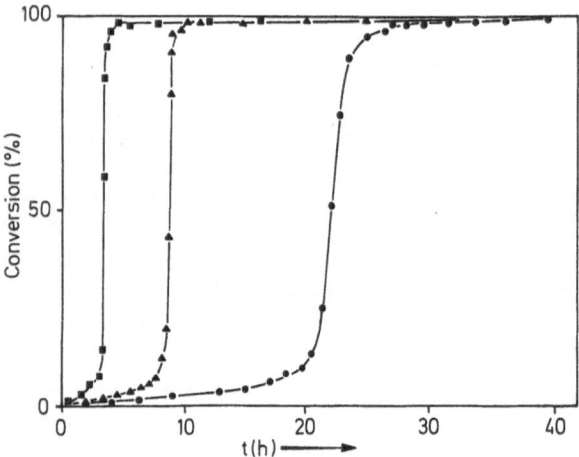

Fig. 10. Time conversion curves for the thermal polymerization of PTS. ●: 60 °C, ▲: 70 °C, ■: 80 °C

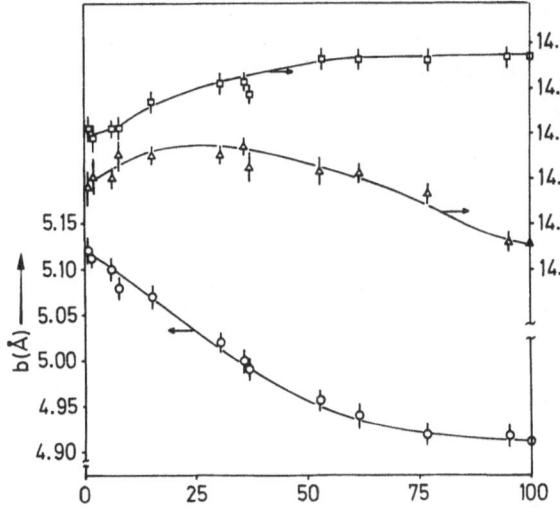

**Fig. 11.** Dependence of the lattice parameters determined at 110 K for PTS on the conversion. $\bigcirc$: b, $\square$: c, $\triangle$: a. The values plotted refer to the high temperature cell

have been used in studies of PTS polymerization. All experimental results indicate that the thermal polymerization of PTS is very close to the limiting case of an ideal homogeneous topochemical reaction. The randomly distributed polymer chains can be regarded as one-dimensional defects giving rise to diffuse sheets in x-ray and neutron scattering experiments which are oriented perpendicular to the chain direction [91-95]. The width of the sheets can be used for a rough estimate of the average chain length as a function of conversion. In the initial stages of polymerization the intensity of diffuse scattering increases. At higher conversions the correlations between chains become increasingly important and are visible as intensity modulations, and finally the diffuse scattering vanishes when complete conversion is reached.

Similar results have been obtained by measuring the Raman spectra of partially polymerized crystals and by other spectroscopic techniques. Here, the vibrational frequency of the polymer chain can be used as a probe for the lattice strain in the vicinity of the dispersed macromolecules [96,97]. It should be noted that monomer and polymer are not strictly isomorphous. The mismatch between monomer stacking and polymer repeat of 0.2 Å per addition step has to be accounted for by the monomer matrix. The raman frequencies shift accordingly to the lattice changes (cf. Fig. 11) in agreement with the random chain distribution.

A polymerizing diacetylene crystal can be considered as a composite material with large differences in the mechanical properties of both components. In such a material the mechanical properties will not only depend on the relative amount of the components but also very strongly on the geometrical arrangements of the structural elements [98-100]. Two limiting cases can be considered: The first model consists of infinite rods of the high modulus material (polymer chain) embedded randomly in the soft monomer matrix and aligned in direction of the strain. This model of a reinforced material has been treated by Voigt [98]. Here, the elastic constant c depends linearly on the volume fraction $v_p$ of the polymer according to:

$$c = (1 - v_p) c_m + v_p c_p \tag{3}$$

$c_m$ and $c_p$ being the elastic constants of the monomer and polymer, respectively.

The other limiting case (Reuss model) of a two component material is a sandwich structure of alternating layers of high and low modulus materials loaded perpendicular to the layer plane [99]. In this case the stress is uniformly distributed within the sample. The resulting modulus is given by:

$$\frac{1}{c} = \frac{1 - v_p}{c_m} + \frac{v_p}{c_p} \tag{4}$$

The two theoretical curves are plotted in Fig. 12 as a function of conversion together with the experimental values of the elastic constant $c_{22}$ of PTS in chain direction which have been obtained by Brillouin scattering [35, 101]. At higher conversions the experimental data are very well represented by the Voigt model. The deviation at low conversions can be explained by the limited length of the polymer chains and have been used to assess the average degree of polymerization.

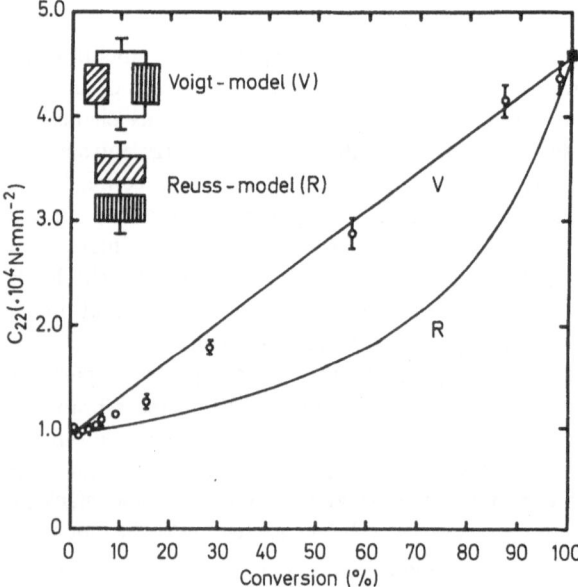

**Fig. 12.** Dependence of the elastic constant $c_{22}$ in chain direction on the conversion in partly polymerized PTS crystals. The black square is the theoretically expected value for the pure polymer. The curves marked (V) and (R) are calculated according to Eq. (3) and (4), respectively

## 4.2 The Crystal Strain Theory

A theory of the polymerization kinetics of diacetylenes has been developed by Baughman [102] on the basis of these findings, i.e. that the polymer is formed in the monomer lattice which it does not match. The resulting strain plays a crucial role in the theory. The following assumptions have been made:
1) In all conversion monomer and polymer units form a solid solution. The separation d in chain direction between lattice sites is conversion dependent.
2) Initiation is constant and conversion dependent.
3) The propagation rate is given by the product of the conversion independent life

time $\tau$ of the active chain end and the conversion dependent kinetic chain length $L_n$:

$$k_p = L_n \tau \tag{5}$$

4) The conversion dependence of $L_n$ can be split into two parts. First, the probability of termination by a previously formed chain in the same stack or lattice defects will increase with conversion. Secondly, the activation energy for the addition of one monomer unit can be approximated by:

$$E_a(X) = D(d(X) - d_p)^2 \tag{6}$$

where D is the force constant in stacking direction and $d(X)$ and $d_p$ the stacking distances at a conversion X and in the pure polymer, respectively.

In a system described by a Voigt model (refer to Eq. 3) Eq. (6) can be rewritten as:

$$E_a(X) = (d_m - d_p)^2 \, c_m \left[ \left( 1 + \frac{X}{1-X} \left( \frac{c_p}{c_m} \right)^{-2} - 1 \right) \right] \tag{7}$$

The theory was used to calculate kinetic curves for the polymerization of PTS deducing the ratio $c_m/c_p$ from the known conversion dependence of the lattice parameters. Time conversion curves normalized with respect to the time necessary to reach 50 percent conversion can be calculated for different values of the lattice mismatch using the crystal strain theory. For PTS a satisfactory fit of the experimental data of the thermal and $\gamma$-ray polymerization can be obtained. However, further studies of the kinetics of the solid-state polymerization of PTS and other monomers provided results which cannot be explained by the theory.

First, the thermal polymerization of PTS is strongly affected by isotope substitution [88, 89]. Deuterated and partially deuterated samples show an unusually large inverse isotope effect, i.e. during the induction period the reaction rate is increased by a factor up to 3.5. In the rapid reaction regime, however, a small normal isotope effect, i.e. a small decrease of the polymerization rate is observed. The magnitude of the isotope effect is largest if the units adjacent to the diacetylene group are substituted. Even substitution of the methylene spacer group with $^{13}C$ has a marked effect on the kinetics. Lattice parameters and thermal activation energies are in all cases uneffected by the isotope substitution. Secondly, the induction period can be suppressed by moderate hydrostatic pressure so that the overall polymerization kinetics becomes nearly first order [103]. The analysis of this effect shows that both the decrease of the induction period and acceleration of the rapid reaction are larger than predicted with the elastic strain theory using the known lattice parameters and elastic constants [104].

Finally, there is experimental evidence that the thermal polymerization of some monomers with mismatches comparable to PTS cannot be described by the elastic strain theory [105, 106].

The polymerization kinetics of diacetylenes is discussed in detail elsewhere [107]. Some of the failures of the theory may be due to the assumptions and approximations made in the calculations. Baughman and Chance have tried to explain some of the above mentioned effects by introducing a chain-terminating impurity [108]. The concentration of this impurity, however, cannot be independently determined and must be fitted according to the experimental results.

## 4.3 Reaction Kinetics and Molecular Weight Distribution

One of the shortcomings of the elastic strain theory which leaves room for further improvements is the fact that assumptions must be made about the individual initiation, propagation and termination steps which cannot be observed independently but are calculated from the overall reaction rate.

A deeper insight into the reaction mechanism may elucidate the conversion dependence of the molecular weight and its distribution. Owing to the extreme insolubility of the better investigated polydiacetylenes like PTS, however, only very limited experimental data have been available until recently from indirect determinations, e.g. from mechanical properties or diffuse scattering of partially polymerized crystals.

Since Patel discovered in 1978 that poly3BCMU ($R = R' = -(CH_2)_3-OCONH-CH_2-COO-(CH_2)_3CH_3$, cf. Table 1) is a readily soluble polymer the study of the solution properties of polydiacetylenes attracted increasing interest [66, 109–115]. With such samples the molecular weight and its distribution can be determined using standard techniques and related to the conversion and other experimental parameters. This analysis has been carried out in great detail for PTS-12

$(\dot R = R' = -(CH_2)_4-OSO_2$—⬡—$CH_3)$     by G. Wenz [115].

The dosage conversion curve for the polymerization of PTS-12 by $^{60}$Co radiation is plotted in Fig. 13 for two temperatures. It is characterized by an induction period followed by a rapid reaction leading to complete conversion which is reached at about 15 Mrad. The onset of the rapid reaction is connected with a phase transition which will be discussed later.

The development of the molecular weight distribution with increasing conversion as determined by gel permeation chromatography is shown in Fig. 14. During the

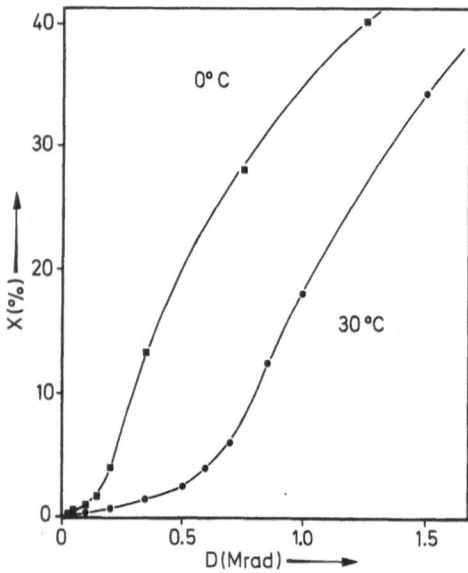

**Fig. 13.** Dosage conversion curves for the polymerization of PTS-12 at 0 °C and 30 °C

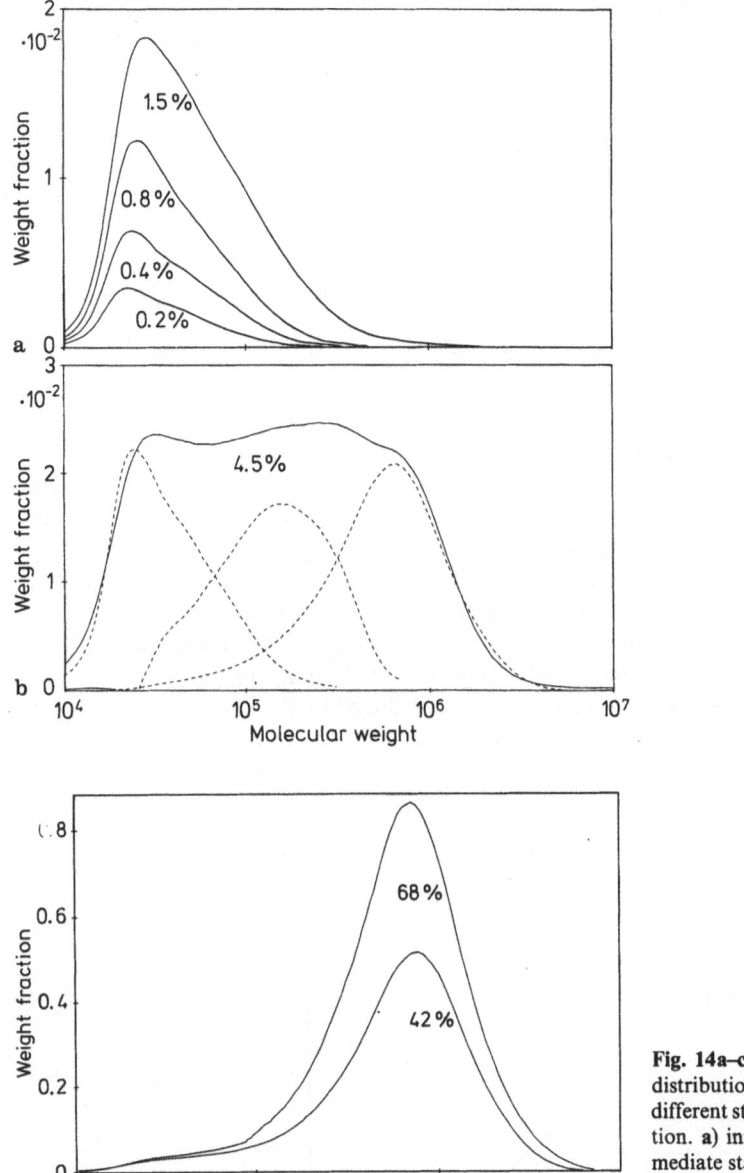

**Fig. 14a–c.** Molecular weight distribution of PTS-12 at three different stages of polymerization. a) initial stage, b) intermediate stage, c) final stage

induction period short chains with an average degree of polymerization $P_n = 60$ are formed. At higher conversions the maximum rapidly shifts to higher values. Three distribution curves with maxima at $P_n = 60$, 200 and 800 can be fitted to the experimental chromatograms. This is evident for the intermediate conversion range. Above approximately 20 percent conversion only the high molecular weight product is formed.

Analysis of the kinetic data shows that the chain initiation rate is constant at $k_i = 6.25 \cdot 10^{-4}$ Mrad$^{-1}$ [G = 1.2 (100 eV)$^{-1}$]. Therefore, the sudden increase of the overall reaction rate in the rapid reaction regime must be attributed to an increase of the kinetic chain length.

It is interesting to note that the short chains formed initially remain intact. This means that there is no re-initiation of "dead" chain ends and no combination of an active chain with a dead polymer present in the same stack. Consequently, two different termination reactions must be considered. In the "free" termination an active chain end is deactivated by a limited life time or some unknown chemical reaction. At higher conversions, however, the "enforced" termination becomes increasingly important; a growing chain is blocked by an already existing polymer. In this kinetic scheme the kinetic chain length L is given by the ratio of the propagation and initiation rates. In Fig. 15 L is compared with the momentaneous degree of polymerization $P_n$ which can be independently determined from difference distributions. Within the experimental error both L and $P_n$ follow the same conversion dependence.

The propagation rate is strongly temperature dependent as already obvious from Fig. 13. In Fig. 16 molecular weight distributions after irradiation with 0.05 Mrad are compared at three different temperatures.

All distribution curves are bimodal with maxima at $P_n = 60$ and 400. At lower temperatures longer chains are formed. Since there is no gradual shift of the maximum with temperature it must be assumed that the chain grows by at least two different active chain ends, the population of which is strongly temperature dependent. The chemical nature of these chain ends cannot be deduced by the kinetic data. However, it seems reasonable to infer that we are dealing with the same carbene and radical intermediates which have been identified in the photopolymerization of diacetylenes at low temperatures by Sixl and coworkers [116].

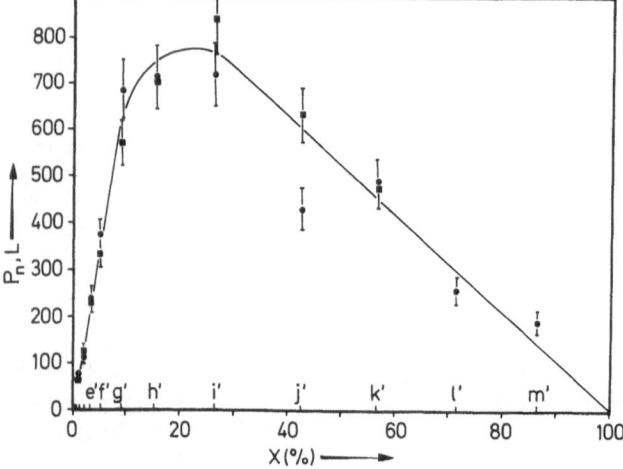

**Fig. 15.** Comparison of the conversion dependence of the kinetic chain length L (●) and the momentaneous degree of polymerization $P_n$ (■) of PTS-12

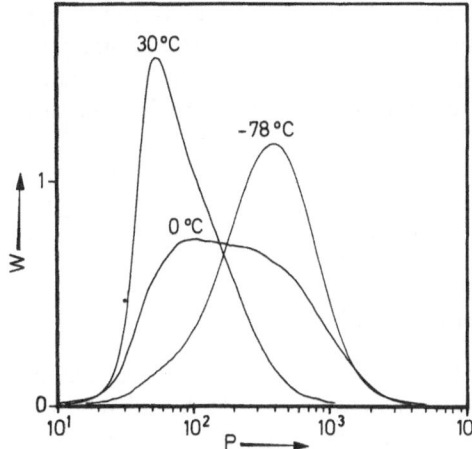

**Fig. 16.** Temperature dependence of the molecular weight distribution of PTS-12 after irradiation with 0.05 Mrad

The following reaction scheme for the polymerization of PTS-12 can be proposed:

$$ \tag{8} $$

It appears that the reaction mechanism and the intermediates involved in the solid-state polymerization of diacetylenes are reasonably well understood. However, experimental results obtained with special monomers should not be generalized. It is not possible to design a monomer with desired properties. Inspection of Table 1 shows that on the basis of the crystallographic data and the monomer packing the absolute reactivity and the polymerization kinetics cannot be quantitatively predicted, e.g. it is not possible, to date, to explain why certain diacetylenes can be polymerized thermally whereas others with equal packing are thermally inactive. A more realistic kinetic model should include the various energy transport processes and the complex side group motions which are connected to the reaction.

## 5 Phase Transitions

The gradual conversion of the monomer crystal into the equivalent polymer crystal can be considered a special type of phase transition. In some cases the topochemical polymerization is accompanied by an additional structural phase transition. This behaviour is most often observed in monomer structures with a comparatively moderate reactivity where only a partial conversion can be achieved. Here, the side group packing is rearranged either spontaneously or by thermal annealing. This process usually leads to phase separation and nucleation of a new, more reactive monomer modification which is subsequently polymerized. The monomer crystal disintegrates in this type of transition although the deposition of the reactive monomer phase may be topotactic, i.e. oriented along certain crystallographic directions leading to a fibrous texture [26, 31, 49].

In some special cases, however, both the polymerization and the side group reorientation are single phase processes. They are of special interest for understanding the dynamics and side group mobility in the solid-state polymerization of diacetylenes.

## 5.1 Conversion Dependent Phase Transition in DCH

1,6-$Di$-(N-$C$arbazolyl)-2,4-$H$exadiyne (DCH) represents such a limiting case for the topochemical polymerization. The packing parameters, $d = 4.55 \text{Å}$ and $\Phi = 60°$ are well outside the range where high reactivity is expected (cf. Fig. 5). Indeed, the polymerization of DCH in the x-ray beam proceeds so slowly that the data collection for the crystal structure analysis could be carried out at room temperature [47].

The dosage conversion curve shown in Fig. 17 shows a distinct induction period. Analysis of the Brillouin scattering has demonstrated that in this range very short chains are formed [117].

At a conversion of approximately 25 percent the crystals undergo a phase transition which is evidenced by a sudden change of all lattice parameters (Fig. 18). Especially the monoclinic angle $\beta$ abruptly changes by 14 degrees during the transition.

Pertinent crystallographic data are summarized in Table 3 and projections of the monomer and polymer structures are shown in Fig. 19 and Fig. 20.

Below the critical conversion the polymer forms a solid solution in the monomer phase. In this state the mismatch is exceptionally large and the polymer chains are contracted by 8 percent. After the transition the situation is reversed and the residual monomer occupies lattice sites within the polymer structure. Here, the packing is much more favourable for the polymerization, which proceeds with large speed.

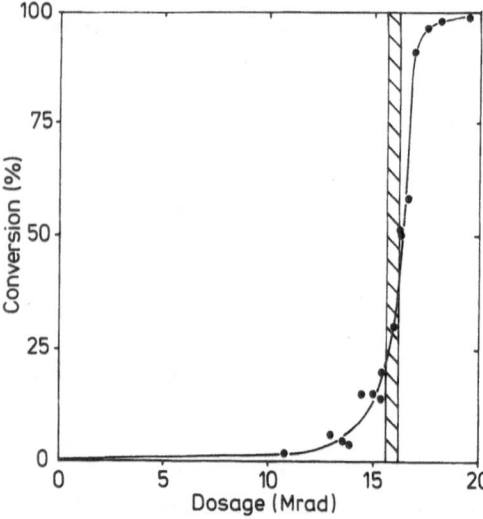

**Fig. 17.** Dosage conversion curve for the polymerization of DCH

In Fig. 18 it can be seen that the carbazole rings retain their stacking distance of 3.35 Å during the transition. This is only possible by a large rotation of the rings about the N-C3 bond. The large shearing of the lattice results from this side group reorientation. It should be emphasized that although we are dealing with a displacive phase transition it proceeds homogeneously throughout the crystal in γ-ray polymerization introducing only little disorder. During the transition the crystal shape changes according to the microscopic changes of unit cell dimensions, i.e. the crystals expand by 8 percent along b and are sheared according to the change of β.

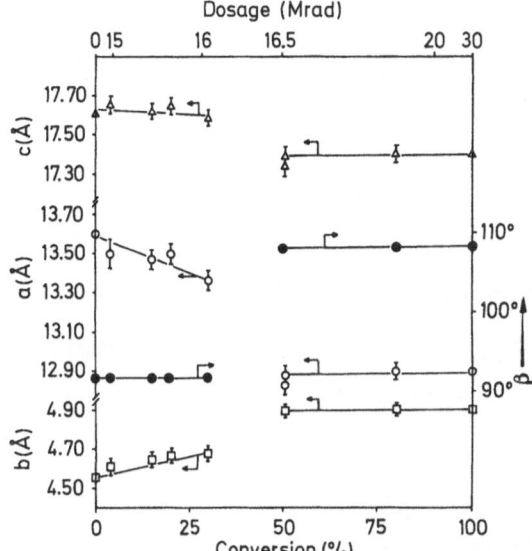

**Fig. 18.** Conversion dependence of the lattice parameters of DCH

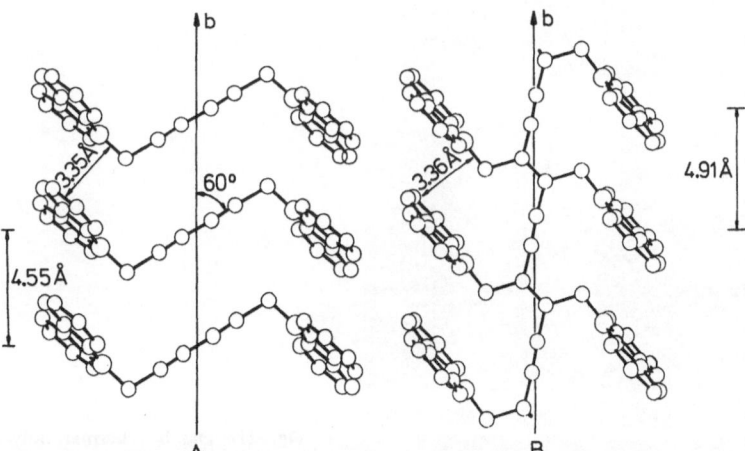

**Fig. 19.** Projection of the DCH monomer and polymer crystal structures on the plane of the polymer chain

It is interesting to note that in contrast to these results the thermal polymerization of DCH always proceeds heterogeneously with nucleation of separate polymer domains. The thermal polymer is polycrystalline with a fibrous texture. Lattice parameters are identical with those of the polymer obtained by irradiation. Observation of thermally polymerizing DCH crystals shows that the reaction starts at crystal

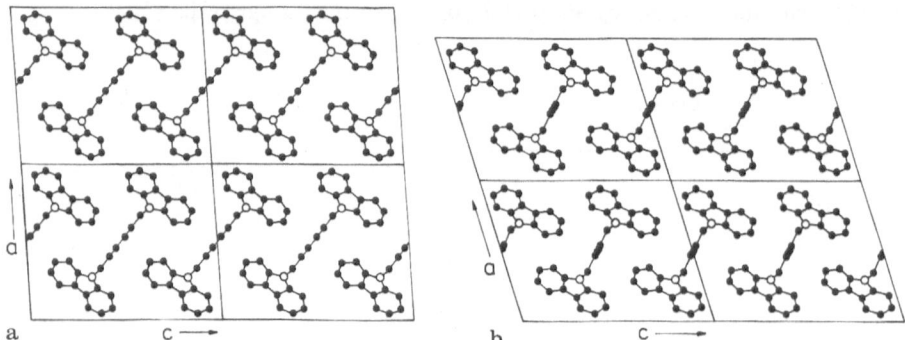

**Fig. 20a and b.** DCH monomer (**a**) and polymer (**b**) crystal structures

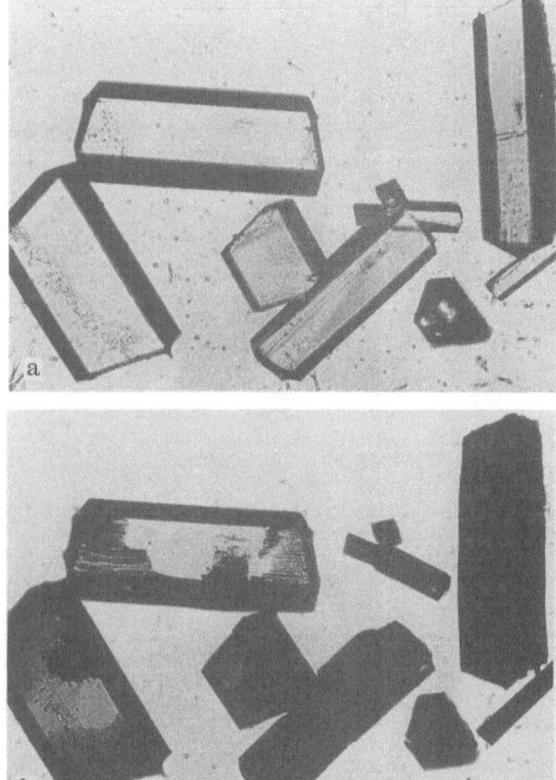

**Fig. 21a and b.** Thermal polymerization of DCH. **a** DCH monomer; **b** after 20 hours at 120 °C

edges or other visible defects. Phase separation then continues by growth of several separate daughter phases inside the monomer crystal. This eventually leads to the destruction of the monomer crystal.

In Fig. 21 DCH crystals are shown before polymerization and at an intermediate conversion. It is typical for the thermal reaction that more perfect monomer crystals require longer reaction times than defect-rich crystals. There is evidence that in the radiation polymerization of DCH the polymer crystal perfection increases with decreasing temperature, i.e., the nucleation process requires a rather high thermal activation energy.

## 5.2 Conversion Dependent Phase Transition in PTS-12

Another phase transition which casts an interesting light on the side group motions is observed during the polymerization of PTS-12 [42]. The onset of the rapid reaction (cf. Fig. 13) is accompanied by this transition. In the monomer the b axis is doubled. The transition can be monitored by a continuous decrease of the intensities of all reflections having odd k indices. The dependence of lattice parameters on conversion is plotted in Fig. 22, pertinent crystallographic data are given in Table 3. The doubling of the monomer unit cell is explained by the fact that in contrast to the polymer chain the monomer unit is not centrosymmetric. Two differently oriented side chains are attached to the reactive triple bond system. This gives rise to a packing arrangement where the centers of the diacetylene groups in neighboring arrays are alternatingly

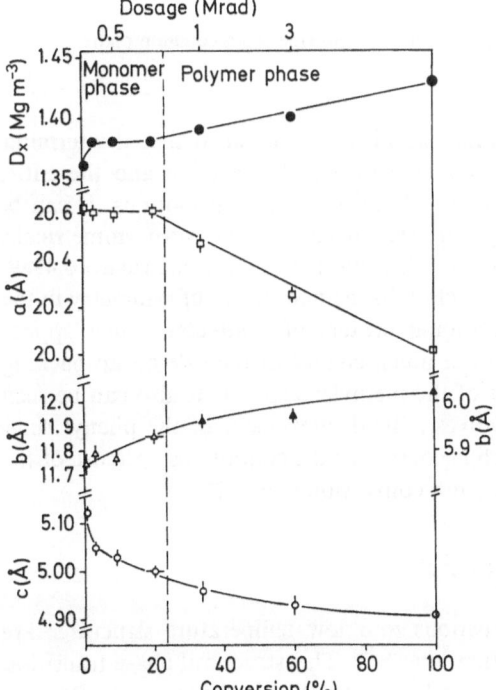

Fig. 22. Conversion dependence of the lattice parameters of PTS-12 determined at 110 K

above and below the lattice site they assume in the polymer structure. This is quite surprising since in addition to the rotation of the diacetylene group a translation of approximately 1 Å is necessary for the reaction which proceeds without phase separation or macroscopic deformation of the crystal.

A projection of the monomer and polymer crystal structures onto a common plane is shown in Fig. 23.

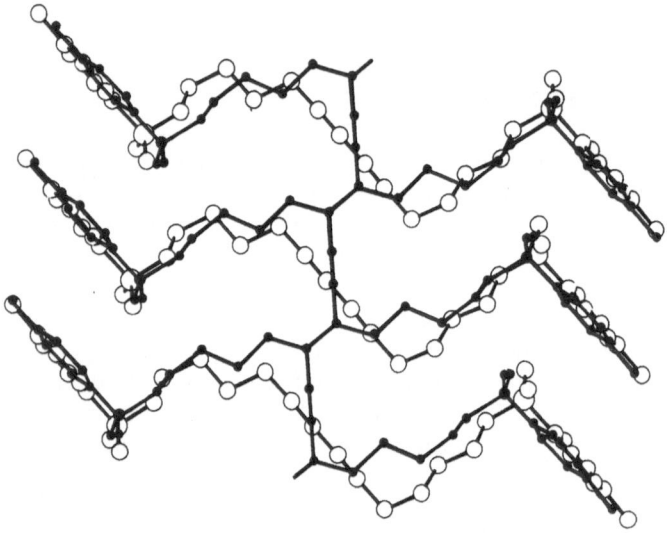

**Fig. 23.** Projection of the structures of PTS-12 monomer and polymer on a common plane

The unexpected behaviour can be understood if it is assumed that the terminal p-toluenesulfonate groups retain their position during the reaction and the entire methylene spacer is included in the rotational and translational motions. It can be seen that indeed the p-toluenesulfonate groups are already pseudo-centrosymmetrically related in the monomer, e.g. the center between the sulfur atoms is located at a = 0.4989, b = 0.2453 and c = 1.0111 which is very close to the new center of symmetry in the polymer structure. Therefore, the additional centers of symmetry which appear at the phase transition are created without much change of the side group packing only by conformational rearrangement of the methylene spacer. It also can be seen in Fig. 23 that this rearrangement involves a small movement of the phenyl rings toward the center of the molecule which accounts for the continuous decrease of the a-axis in the polymer phase with increasing conversion (Fig. 22).

## 5.3 The Low Temperature Phase of PTS

In PTS monomers and polymers transitions into low temperature structures are observed which have attracted much attention [36, 79]. This structural phase transition is characterized by a doubling of the unit cell as schematically shown in Fig. 24.

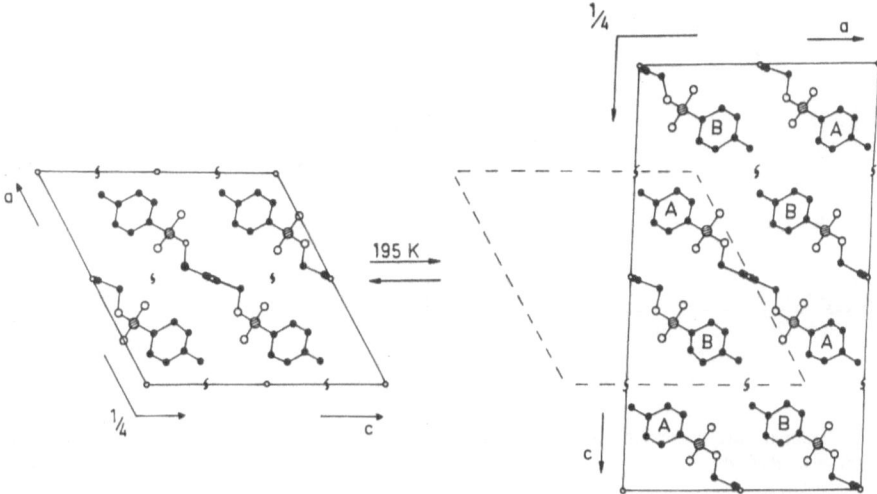

**Fig. 24.** Projections of the crystal structure of PTS polymer above and below the phase transition. A and B represent different side group orientations

The transition is caused by the torsion of the side groups in a manner resembling the formation of periodic stacking faults. All phenyl rings belonging to one row of molecules oriented along the (102) plane of the high temperature structure rotate by about 8 degrees in one direction, the side groups in the neighboring rows in the opposite direction. This creates crystallographically different "A" and "B" molecules in the unit cell.

Since there is only little side group reorientation the transition is connected with small anomalies of the temperature dependence of the lattice parameters (Fig. 25).

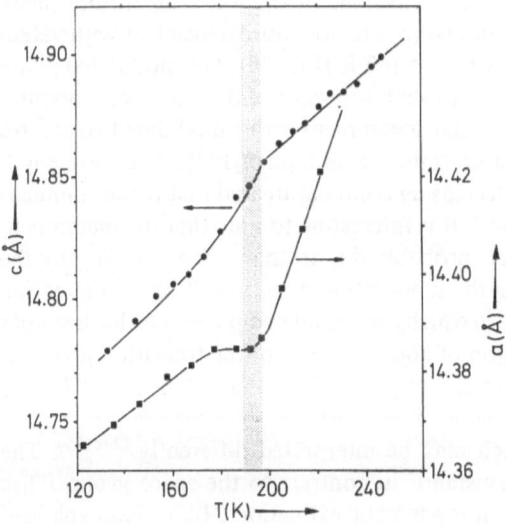

**Fig. 25.** Temperature dependence of the a (■) and c (●) lattice parameters of PTS polymer. The superstructure reflections appear in the shaded temperature range

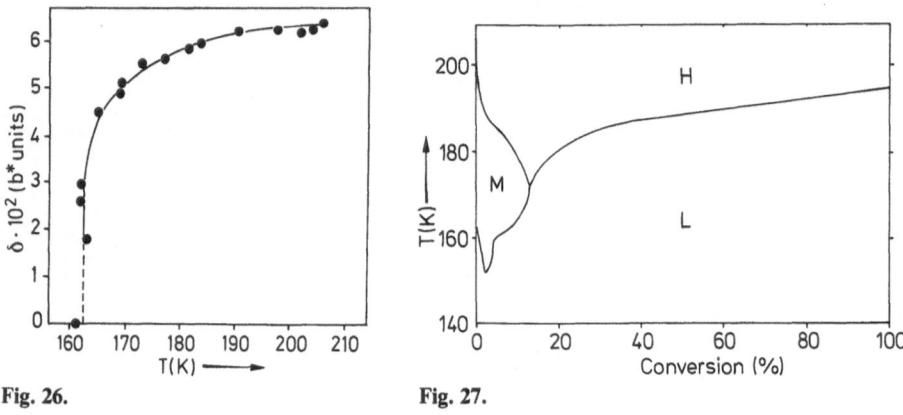

Fig. 26.                                              Fig. 27.

**Fig. 26.** Temperature dependence of the splitting δ of the satellites in the modulated PTS structure according to Ref. 95

**Fig. 27.** Phase diagram of PTS according to Ref. 81. H: high temperature phase, M: modulated phase, L: low temperature phase

In a small temperature range indicated in Fig. 25 additional reflections appear and all bands in the spectra begin to split into two [118]. It is interesting to note that within the experimental error the bond lengths and angles of the polymer chain stay constant in the whole temperature range and are equal in the two independent chains. The comparatively large splitting of the optical absorption of the polymer backbone ($\sim 500$ cm$^{-1}$) must therefore be attributed to the small side group reorientation.

In PTS monomer and partly polymerized crystals similar transitions are observed. However, in the monomer an additional incommensurable modulated structure is present as a precursor of the low temperature structure [81, 92–95, 119, 120]. The modulated phase is characterized by satellite reflections which appear at 0.5 a*, δb*, 0 in reciprocal space, i.e. the superstructure reflections of the low temperature phase are split into two. The splitting δ in b* direction is temperature dependent with values for δ ranging from 0.062 b* at 206 K to zero at 163 K (Fig. 26). The modulated phase can be described involving an occupation probability wave of the side group orientation (molecules A and B in Fig. 24). Similar incommensurable modulated structures have been observed in other molecular crystals, e.g. in biphenyl [121]. The wavelength of the modulation δ$^{-1}$ increases with decreasing temperature and finally the commensurate low temperature phase "locks in". It is interesting to note that the incommensurable phase is absent in the polymer, probably due to the strong coupling by the polymer chain which is oriented along the modulation direction. With an increasing polymer content the temperature range in which the modulated phase is stable becomes increasingly smaller. Above a conversion of about 13 percent the transition proceeds directly from the high to the low temperature phase. The phase diagram of PTS is schematically shown in Fig. 27. In the region of the low temperature transitions a pyroelectric effect has been observed which may be interpreted differently [122, 123]. The apparent lack of a center of symmetry stands in contrast to the space group P2$_1$/c which is unambiguously determined by the systematic extinctions. Bloor has explained these findings by local pertubations or macroscopic deformation of the crystal coming from the internal strain produced during polymerization [123].

## 5.4 The Low Temperature Phase of DCH

At 142 K DCH monomer undergoes a first order phase transition into an unreactive low temperature structure. The transition is accompanied by a sudden decrease of the stacking axis b. The crystallographic data are included in Table 3, projections of the two crystal structures are shown in Fig. 28 in comparison.

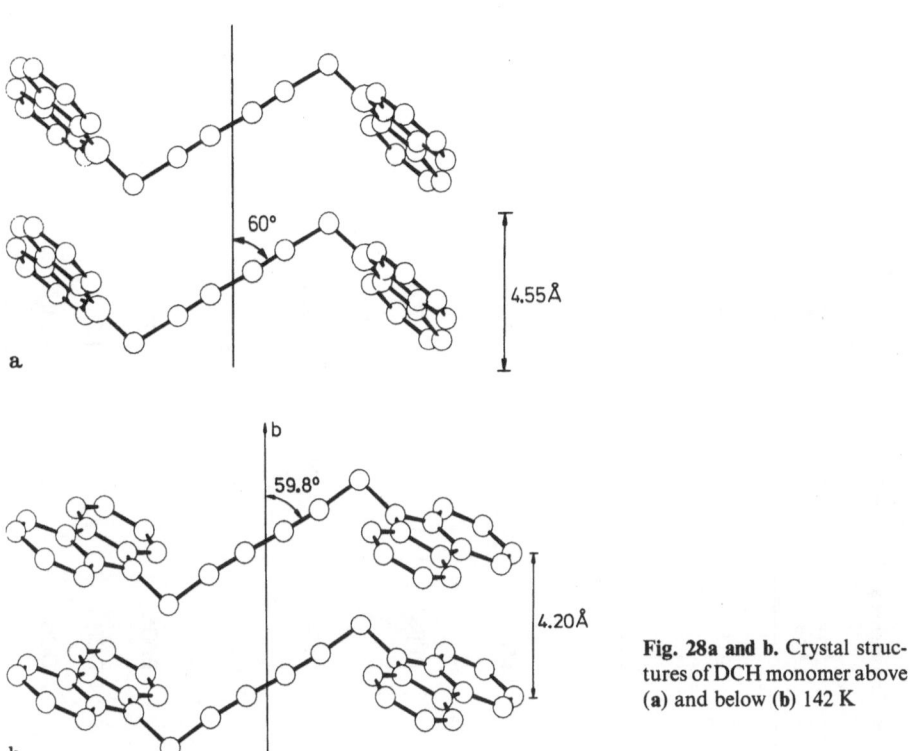

Fig. 28a and b. Crystal structures of DCH monomer above (a) and below (b) 142 K

The phases are readily identified by the colour of polymer traces present in the crystal, blue above and red below 142 K [124]. Sharp phase boundaries between the phases have been identified by Bloor et al. as the (031) and (041) crystallographic planes of the two structures. Larger crystals tend to fracture during the transition due to nucleation of several low temperature phases at crystal defects. The large change in crystal density ($\sim$4%) renders the transition temperature sensitive to hydrostatic pressure [125, 126]. At a pressure of about $3 \cdot 10^2$ MPa the critical temperature is shifted to room temperature. This low temperature structure is not observed in DCH polymer or the anthracene substituted analogs DAH and ACH.

**Table 4.** Observed bond lengths of polydiacetylene chains and [structures] of butatriene model compounds

| R | Abbr. | T/K | A/Å | B/Å | C/Å | Ref. |
|---|---|---|---|---|---|---|
| —CH$_2$—OSO$_2$—⟨CH$_3$⟩ | PTS | 295 | 1.191(4) | 1.428(4) | 1.356(4) | 78) |
| —CH$_2$—OSO$_2$—⟨CH$_3$⟩ | PTS | 120 | 1.19(2) | 1.43(2) | 1.36(2) | 79) |
| —CH$_2$—OSO$_2$—⟨OCH$_3$⟩ | MBS | 295 | 1.195(5) | 1.424(3) | 1.364(5) | 127) |
| —CH$_2$—OSO$_2$—⟨F⟩ | PFBS | 295 | 1.17(2) | 1.39(3) | 1.42(3) | 38) |
| —(CH$_2$)$_4$—OSO$_2$—⟨CH$_3$⟩ | PTS-12 | 295 | 1.15(2) | 1.41(1) | 1.41(2) | 42) |
| ⟨CO(CH$_2$)$_3$CO⟩ | BPG* | 295 | 1.29 | 1.38 | 1.42 | 43, 44) |
| —CH$_2$—Cz | DCH | 295 | 1.21 | 1.44 | 1.33 | 84) |
| —CH$_2$—Cz | DCH | 295 | 1.23(1) | 1.42(1) | 1.38(1) | 72, 85) |

| Structure | Name | Temp. | | | | Ref. |
|---|---|---|---|---|---|---|
| —CH₂—OCONH— | HDU-1 | 295 | 1.21 | 1.42 | 1.36 | 128) |
| —(CH₂)₄—OCONH— | TCDU-1 | 295 | 1.17(2) | 1.38(2) | 1.46(2) | 83) |
| —CH₂—O—(3,4-dinitrophenyl) | DNP | 295 | 1.205(7) | 1.410(5) | 1.471(9) | 58) |
| —CH₂—N(Ph)₂ | THD | 295 | 1.205(2) | 1.426(3) 1.427(3) | 1.359(3) 1.360(3) | 71) |
| | Butatriene structure (theor.) Acetylene structure (theor.) | | 1.28 1.21 | 1.32 1.43 | 1.46 1.34 | 128) 128) |
| (phenyl) | butatriene (exp.) | 100 | 1.259 | 1.348 | 1.478 | 129) |
| (cycl.) H₃C CH₃ / H₃C CH₃ | butatriene (exp.) | 295 | 1.261 | 1.332 | 1.547 | 130) |

* Solid solution, 35% polymer in monomer

## 6 Structure of the Polydiacetylene Chain

Two different reaction mechanisms have been postulated for the topochemical polymerization of diacetylenes involving diradical or carbene chain ends [116]. The first mechanism leads to a butatriene structure (I), the latter to the acetylene structure (II) of polydiacetylene chains.

$$(9)$$

The elucidation of the reaction mechanism and the identification of the reactive intermediates has been a matter of debate and extensive studies in several laboratories; it will be covered in other reviews [116]. Sixl et al. have been able to demonstrate that the energetically favoured diradicals giving rise to the butatriene structure (I) are only observed for oligomers with comparatively short chain length (n < 7). In longer chains the higher energy of the carbene intermediates is overcompensated by the lower energy of the resonance structure (II) and in later stages of the reaction carbenes and dicarbenes can be identified to be the reactive chain ends. Therefore, the acetylene structure (II) is expected for the polydiacetylene chain.

The determination of bond lengths and angles of the polymer backbone is only possible in cases where the polymerization can be carried out to quantitative conversion with retention of the crystal perfection. It has been mentioned above that this condition is fulfilled in varying degrees for different reactive diacetylenes.

The experimental data for various polydiacetylenes and for two model compounds for the butatriene structure are compiled in Table 4.

It should be noted that the values given in Table 4 reflect both the different qualities of the crystal structure analyses and of the crystals. The bond lengths are not corrected for anisotropic thermal vibrations. From the differences found in two independent structure analyses of DCH polymer it can be assumed that in some cases the standard deviations given may be underestimated. In all cases the quality of the analyses does not allow the determination of the electron density distribution along the polymer chain which has been possible for the two model compounds and for the resonance structure (I) [129, 130].

Within these limits the polydiacetylene chain seems to be best represented by the acetylene structure (II). This is especially true for those polymers (PTS, HDU-1, THD) which are obtained thermally under mild conditions. In the other cases deviations from the expected values can be primarily explained by defects which are introduced by γ-irradiation and in other cases by pertubations of the electron density of the polymer chain by end groups and residual monomer units. Examples for this behaviour are TCDU-1 and BPG-1 which have been claimed to represent strong admixtures of the resonance structure (I). As it has been mentioned above the polymerization of TCDU-1 is accompanied by unusually large side group rotations giving rise to large lattice changes (cf. Table 3). This introduces some disorder which is evidenced by large anisotropic thermal parameters and unusual bond lengths in the side chain [83].

The second example, BPG-1, is a case of limited reactivity. The crystal structure analysis was carried out using a crystal containing only 35 percent polymer [44]. The quality of the data clearly does not allow to draw conclusions on the electronic structure of the polymer chain.

A similar argument can also be used for DNP which shows an unusually large bond length C2–C2' ("C" in Table 4). Here, the thermal polymerization is followed by a second thermally activated process leading to the destruction of the crystal [58]. This has been interpreted as a disruptive phase transition [93]. The unusual bond lengths can be attributed to defects introduced by the second reaction which competes with polymerization.

Resonant raman spectroscopy has proved to be another valuable tool for the study of the structure of the polydiacetylene chain. Due to the resonance enhancement the spectra are compared to greatly simplified, infrared spectra and show as principle feature only the in-plane modes of the polymer chain. The correlation of the $C \equiv C$ and $C = C$ stretching modes and their temperature dependence have been interpreted as resonances between the mesomeric structures (I) and (II) [131, 132]. However, a model using simple anharmonic force constants for the acetylene structure (II) is in good agreement with the experiment, e.g. the temperature and pressure dependence of the vibration frequency and the mechanical properties [133–135].

Many polydiacetylenes show drastic colour changes in the solid-state or in solution when the temperature or the solvent is changed [109–111, 114, 136–140]. This "red-to-blue" transition which is shown for one example in Fig. 29 has been interpreted in terms of different resonance contributions or changes of planarity.

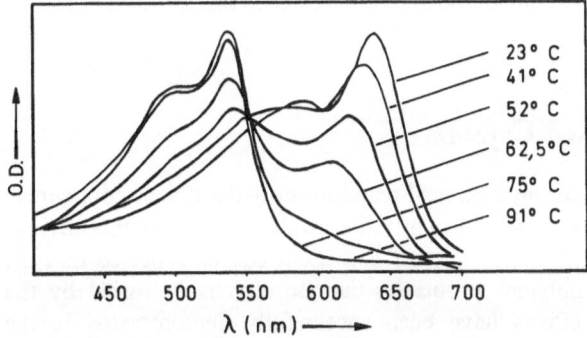

Fig. 29. Red-to-blue transition in a polydiacetylene

However, it should be emphasized at this point that despite the great effort devoted to the elucidation of this effect, which has been observed very early, to date there is no experiment which unambiguously explains the nature of the colour changes.

Two models for the shape of the polydiacetylene chain in solution are schematically presented in Fig. 30.

The first model (Kuhn chain) [141] is built up by planar segments of limited conjugation length which are separated by defects, e.g. cis double bonds. The second concept of a "worm-like" chain (Porod-Kratky chain) [142, 143] visualizes a continuous curvature of the chain skeleton. In this model the chain stiffness is characterized by the average angle between two segments or the persistence length $l_{pers}$. Recent studies of the solution properties of PTS-12 have shown that all data can be readily discussed in terms of the concept of "worm-like" chains.

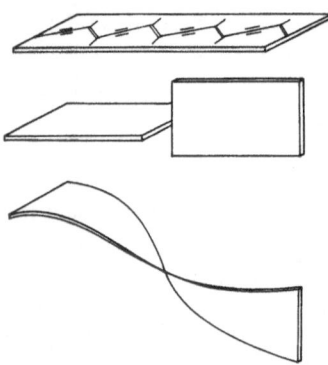

**Fig. 30.** Schematic representation of the shape of polydiacetylene chains. Top: planar, fully conjugated chain, middle: Kuhn model, bottom: worm-like chain

The dramatic colour changes which are observed in solutions of certain polydiacetylenes, e.g. poly3BCMU or poly4BCMU, when the solvent to non-solvent ratio or the temperature is changed, have been interpreted as single-chain coil-to-rod transition [109–111, 144, 145]. However, this is still a matter of debate and continuing experiments. There is experimental evidence that this transition is connected with an aggregation process [114, 146].

In conclusion it must be admitted that the spectral changes of polydiacetylene chains in various environments, which are intimately coupled to the electronic structure of the backbone, are still not fully understood and remain one of the unsolved problems in this field.

# 7 Polymerization in Mixed Crystals

Owing to the special principles of topochemical reactions polydiacetylene copolymers can only be obtained if the comonomers form mixed crystals. Apart from this obvious preparative aspect mixed crystals are of interest because it can be expected that the monomer reactivity and the polymer properties can be effectively varied by the crystal composition. Similar effects have been successfully demonstrated in the "isostructural doping" of organic charge-transfer crystals [147]. In addition, it can be imagined that comonomer units introduced into the host lattice can be regarded as well defined defect sites. These could be used as probes in order to obtain further insight into the complex energy transfer processes during the reaction.

A necessary and sufficient condition for the formation of substitutional solid solutions of organic molecules is similarity of shape and size of the component molecules. For the formation of a continuous series of solid solutions the crystal structures of the pure components must be isomorphous [148]. Due to the rather irregular shape of organic molecules the principle of close packing leads to structures of low symmetry so that the latter requirement is not often fulfilled. Several diacetylenes which were found to form mixed crystals are given in Table 5. A large number of

**Table 5.** Diacetylene monomers forming mixed crystals
A: anthryl group, Cz: carbazolyl group

| Monomer 1 | Monomer 2 R | R' | Ref. |
|---|---|---|---|
| PTS | $-CH_2OSO_2-$[benzene]$-F$ | $= R$ | 38) |
| PTS | $-CH_2OSO_2-$[benzene]$-Cl$ | $= R$ | 149,150) |
| PTS | $-CH_2OSO_2-$[benzene]$-Br$ | $= R$ | 150) |
| PTS | $-CD_2OSO_2-$[benzene D]$-CD_3$ | $= R$ | 88,89) |
| PTS | $-CD_2OSO_2-$[benzene]$-CH_3$ | $= R$ | 88,89) |
| PTS | $-CH_2OSO_2-$[benzene D]$-CD_3$ | $= R$ | 88,89) |
| PTS | $^{13}-CH_2OSO_2-$[benzene]$-CH_3$ | $= R$ | 88,89) |
| PTS | $-CH_2OSO_2-$[benzene]$-OCH_3$ | $= R$ | 151) |
| PTS | $-CH_2OSO_2-$[naphthalene] | $= R$ | 151) |
| DCH | $-CH_2A$ | $= R$ | 48) |
| DCH | $-CH_2A$ | $-CH_2Cz$ | 48) |
| HDU-1 | $-CH_2OCONH-$[thiophene] | $= R$ | 152) |
| HDU-1 | $-CH_2OCONH-$[benzene]$-Cl$ | $= R$ | 153) |

comonomers forming mixed crystals with PTS have been obtained by chemical modification of the p-toluenesulfonate side group. Of these only the various deuterated and $^{13}C$ labelled compounds as well as PFBS form mixed crystals over the whole concentration range. The isotope labelled compounds form ideal solid solutions in the PTS matrix and consequently all properties, e.g. the kinetic isotope effects change accordingly to the comonomer content.

PFBS $(R = -CH_2-OSO_2-\langle\bigcirc\rangle-F)$ crystallizes in a structure which is isomorphous

to PTS and exhibits virtually identical packing parameters. However, thermal polymerization proceeds much slower and by changing the crystal composition the reactivity can be changed by a factor of 8 [38, 154]. Time conversion curves are shown in Fig. 31. The PFBS comonomer acts as a lattice site where the chain termina-

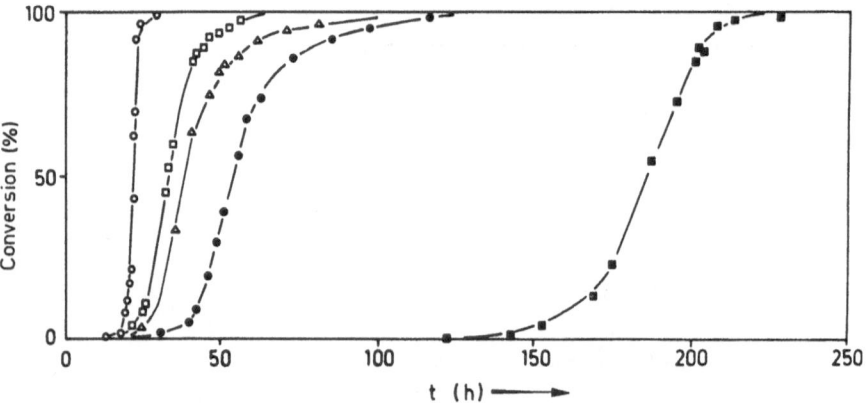

**Fig. 31.** Time conversion curves for the thermal polymerization of PTS-PFBS mixed crystals at 60 °C. ○: PTS, ■: PFBS, □: 20 mole % PFBS, △: 50 mole % PFBS, ●: 62 mole % PFBS

tion probability is greatly enhanced. Since the rapid reaction regime has much higher kinetic chain lengths it is affected by much smaller admixture of the comonomer than the reaction in the induction period. The kinetic data are plotted in Fig. 32.

The other comonomers for PTS given in Table 5 form mixed crystals only in a limited concentration range. In contrast to PBS the chloro and bromo derivatives PCS and PBS form completely unreactive triclinic structures (cf. Table 1). However, by crystallization of these monomers at very high supersaturations crystals of metastable reactive modifications are obtained which have been shown to be isomorphous to PTS [150]. Owing to the crystallization conditions these crystals are unfortunately not suited for detailed structural investigations. The metastable phases melt up to 50 degrees below the stable modifications which spontaneously recrystallize from the melt.

When PCS and PBS are cocrystallized with PTS two sorts of mixed crystals are formed simultaneously, i.e. active crystals in which up to 20 percent of PCS can be incorporated in the PTS lattice and inactive crystals where PTS is dissolved in the comonomer structure. The cocrystallization diagram for the system PTS-PCS is shown in Fig. 33.

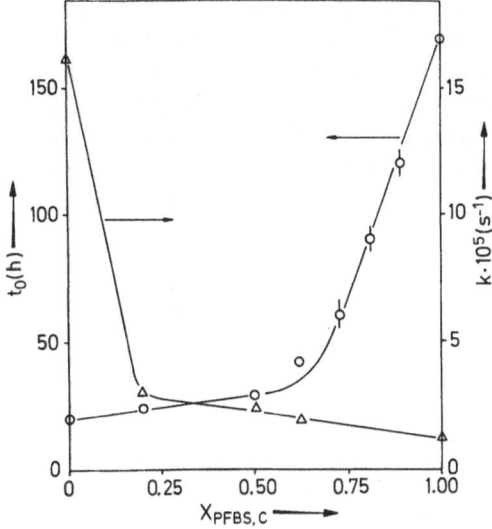

Fig. 32. Dependence of the induction period $t_0$ and the first order rate constant k of the rapid reaction for the data plotted in Fig. 31

The dependence of crystal reactivity is rather complicated. At small admixtures ($x_{PCS} < 0.05$) the induction period is decreased but at higher concentrations substantially increased.

It is interesting to note that the transition into the low temperature phases is inhibited by the replacement of a relatively small number of terminal methyl groups. It seems that the phase transition is mediated by interactions of the methyl groups. Replacement of about 10 percent of the methyl groups by fluorine is sufficient to suppress the transition below 110 K although the free volume associated with the methyl group rotation is retained [38]. In the more closely packed polymer this effect is even more

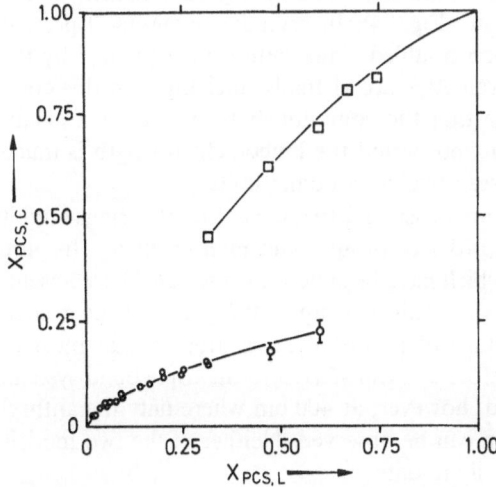

Fig. 33. Cocrystallization diagram for the crystallization of PTS-PCS mixed crystals. $x_{PCS,L}$, $x_{PCS,C}$ are mole fractions of PCS in the acetone solution and in the mixed crystals

pronounced. Pure PFBS monomer and polymer show no phase transition down to 4.2 K [155, 156]. The other diacetylene monomer which is known to form substitutional mixed crystals over the whole concentration range is DCH. The anthracene rings in the comonomer DAH and ACH are able to replace the carbazolyl groups. This is demonstrated by the fact that ACH forms an order-disorder structure which is isomorphous to DCH and in which both side groups statistically occupy the same lattice sites. Owing to the closer packing DAH and ACH are completely inactive (cf. Table 1).

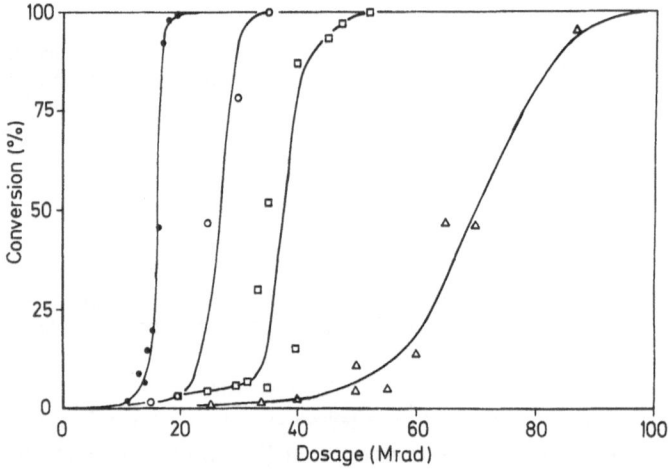

**Fig. 34.** Dosage conversion curves for the γ-ray polymerization of DCH-ACH mixed crystals. ●: DCH, ○: 1.1 mole % ACH, □: 2.2 mole % ACH, △: 5.5 mole % ACH

The admixture of small amounts of anthracene containing comonomers has a drastic effect on the polymerization rate (Fig. 34). In crystals containing 2 percent ACH the induction period is more than doubled. This cannot be explained by the monomer packing since the lattice parameters are virtually unchanged in this composition range. It is also not probable that the comonomer would act as a chain-terminating impurity since in the induction period the kinetic chain length is much smaller than the average distance between two comonomer units.

Two explanations for this behaviour are possible. First, the anthracene rings inhibit phase transition which is necessary in order to bring about high reactivity. In pure DAH and ACH the phase transitions which have been described for DCH are absent.

Secondly, it has been proposed that the anthryl groups act as traps for the excited states which are formed in the first stage of photopolymerization so that most of the energy is emitted as fluorescence before it is transferred to the triple bond system.

If these mixed crystals are irradiated, however, at 400 nm where only the anthryl groups absorb a small photosensation can be observed. Neither of the two models have been fully established experimentally to date.

# 8 References

1. Baeyer, A., Landsberg, L.: Ber. dt. chem. Ges. *15*, 61 (1884)
2. Baeyer, A., Bloem, L.: Ber. dt. chem. Ges. *17*, 964 (1886)
3. Castille, A.: Bull. Acad. Roy. Med. Belg. *6*, 152 (1941)
4. Dunitz, J. D., Robinson, J. M.: J. Chem. Soc. *1947*, 1145
5. Bowden, K., Heilbron, I., Jones, E. R. H., Sargeant, K. H.: J. Chem. Soc. *1947*, 1579
6. Armitage, J. B. A., Cook, C. L., Entwistle, N., Jones, E. R. H., Whiting, M. C.: J. Chem. Soc. *1952*, 1998
7. Bohlmann, F.: Ber. dt. chem. Ges. *84*, 785 (1951)
8. Black, H. K., Weedon, B. C. L.: J. Chem. Soc. *1953*, 1785
9. Seher, A.: Liebigs Ann. Chem. *589*, 222 (1954)
10. Bohlmann, F.: Angew. Chem. *69*, 82 (1957)
11. Hirshfeld, F. L., Schmidt, G. M. J.: J. Polym. Sci. A *2*, 2181 (1964)
12. Wegner, G.: Z. Naturforsch. (b) *24*, 824 (1969)
13. Schmidt, G. M. J.: Solid state photochemistry. Weinheim: Verlag Chemie 1976
14. Liebermann, C.: Ber. dt. chem. Ges. *22*, 124 (1889); *22*, 782 (1889)
15. Wegner, G.: Recent progress in the chemistry and physics of poly (diacetylenes). In: Molecular metals. Hatfield, W. E. (ed.). New York: Plenum Press 1979, pp. pp. 209–242
16. Bloor, D.: Springer Lect. Notes in Phys. *113*, 14 (1980)
17. Baughman, R. H., Chance, R. R.: Ann. N.Y. Acad. Sci. *313*, 705 (1978)
18. Wegner, G.: Organic linear polymers with conjugated double bonds. In: Chemistry and physics of one-dimensional metals. Keller, H. J. (ed.). New York: Plenum Press 1977, pp. 297–314
19. Wegner, G.: Pure and Appl. Chem. *49*, 443 (1977)
20. Wegner, G.: Faraday Disc. *68*, 494 (1980)
21. Enkelmann, V.: Springer Lect. Notes in Phys. *113*, 1 (1980)
22. Baughman, R. H.: J. Polym. Sci., Polym. Phys. Ed. *12*, 1511 (1974)
23. Baughman, R. H., Yee, K. C.: J. Polym. Sci., Macromol. Rev. *13*, 219 (1978)
24. Enkelmann, V.: Coll. and Polym. Sci. *256*, 893 (1978)
25. Bloor, D.: The polymerization of disubstituted diacetylene crystals. In: Developments in crystalline polymers. Bassett, D. C. (ed.). Barking: Applied Science Publishers 1982, pp. 151–193
26. Wegner, G., Fischer, E. W., Munoz-Escalona, A.: Makromol. Chem. Suppl. *1*, 521 (1975)
27. Penzien, K., Schmidt, G. M. J.: Angew. Chem. Int. Ed. *8*, 608 (1969)
28. Elgavi, A., Green, B. S., Schmidt, G. M. J.: J. Amer. Chem. Soc. *95*, 2058 (1973)
29. Green, B. S., Lahav, M., Rabinovich, D.: Acc. Chem. Res. *12*, 191 (1979)
30. Baughman, R. H., Chance, R. R., Cohen, M. J.: J. Chem. Phys. *64*, 1869 (1976)
31. Kaiser, J., Wegner, G., Fischer, E. W.: Israel J. Chem. *10*, 157 (1972)
32. Meyer, W., Lieser, G., Wegner, G.: Makromol. Chem. *178*, 631 (1977); J. Polym. Sci., Polym. Phys. Ed. *16*, 1365 (1978)
33. Braun, H. G., Wegner, G.: Mol. Cryst. Liq. Cryst. *96*, 121 (1983); Braun, H. G., Enkelmann, V.: Acta Cryst. C, submitted
34. Wegner, G., Enkelmann, V.: Angew. Chem. *89*, 432 (1977)
35. Enkelmann, V., Leyrer, R. J., Wegner, G.: Makromol. Chem. *180*, 1787 (1979)
36. Enkelmann, V., Wegner, G.: Makromol. Chem. *178*, 635 (1977)
37. Mayerle, J. J., Clarke, T. C.: Acta Cryst. B *34*, 143 (1978)
38. Enkelmann, V.: Makromol. Chem. *184*, 1945 (1983)
39. Fisher, D. A., Ando, D. J., Bloor, D., Hursthouse, M. B.: Acta Cryst. B *35*, 2075 (1979)
40. Williams, R. L., Ando, D. J., Bloor, D., Hursthouse, M. B.: Acta Cryst. B *35*, 2075 (1979)
41. Williams, R. L., Ando, D. J., Bloor, D., Hursthouse, M. B., Motevalli, M.: Acta Cryst. B *36*, 2155 (1980)
42. Siegel, D., Sixl, H., Enkelmann, V., Wenz, G.: Chem. Phys. *72*, 201 (1982)
43. Lando, J. B., Day, D., Enkelmann, V.: Polym. Symp. *65*, 73 (1978)
44. Day, D., Lando, J. B.: J. Polym. Sci., Polym. Phys. Ed. *16*, 1009 (1978)
45. Enkelmann, V., Graf, H. J.: Acta Cryst. B *34*, 3715 (1978)
46. Mayerle, J. J., Flandera, M. A.: Acta Cryst. B *34*, 1374 (1978)
47. Enkelmann, V., Schleier, G., Wegner, G., Eichele, H., Schwoerer, M.: Chem. Phys. Lett. *52*, 314 (1977)

48. Enkelmann, V., Schleier, G., Eichele, H.: J. Mater. Sci. *17*, 533 (1982)
49. Kaiser, J.: Dissertation. Mainz: 1972
50. Enkelmann, V.: J. Chem. Res. (S) *1981*, 344; J. Chem. Res. (M) *1981*, 3901
51. Gross, H., Sixl, H., Kröhnke, C., Enkelmann, V.: Chem. Phys. *45*, 15 (1980)
52. Patel, G. N., Duesler, E. N., Curtin, D. Y., Paul, I. C.: J. Amer. Chem. Soc. *102*, 461 (1980)
53. Hädicke, E., Penzien, K., Schnell, H. W.: Angew. Chem. *83*, 1024 (1971)
54. Fisher, D. A., Ando, D. J., Batchelder, D. N., Hursthouse, M. B.: Acta Cryst. B *34*, 3799 (1978)
55. Fisher, D. A., Batchelder, D. N., Hursthouse, M. B.: Acta Cryst. B *34*, 2365 (1978)
56. Morosin, B., Harrah, L.: Acta Cryst. B *33*, 1760 (1977)
57. Hanson, A. W.: Acta Cryst. B *31*, 831 (1975)
58. McGhie, A. R., Lipscomb, G. F., Garito, A. F., Desai, K. N., Kalyanaraman, P. S.: Makromol. Chem. *182*, 965 (1981)
59. Wibenga, E. H.: Z. Kristallogr. *102*, 193 (1940)
60. Mayerle, J. J., Clarke, T. C., Bredfeldt, K.: Acta Cryst. B *35*, 1519 (1979)
61. Jeffrey, G. A., Rollet, J. S.: Proc. Roy. Soc. A *213*, 86 (1952)
62. Brouty, C., Spinat, P., Whuler, A.: Acta Cryst. B *36*, 2624 (1980)
63. Enkelmann, V., Schleier, G.: Acta Cryst. C, in press
64. Enkelmann, V., Schleier, G.: Acta Cryst. C, in press
65. Williams, R. L., Ando, D. J., Bloor, D., Motevalli, M., Hursthouse, M. B.: Acta Cryst. B *38*, 2078 (1982)
66. Enkelmann, V., Wenz, G., Müller, M. A., Schmidt, M., Wegner, G.: Mol. Cryst. Liq. Cryst. in press
67. Spinat, P., Whuler, A., Brouty, C.: Acta Cryst. C *39*, 1084 (1983)
68. Brouty, C., Spinat, P., Whuler, A.: Acta Cryst. C *39*, 594 (1983)
69. Coles, B. F., Hitchcook, P. B., Walton, D. R. M.: J. Chem. Soc. Dalton *1975*, 442
70. Enkelmann, V., Schleier, G.: unpublished results
71. Enkelmann, V., Schleier, G.: Acta Cryst. B *36*, 1954 (1980)
72. Enkelmann, V., Kröhnke, C.: unpublished results
73. Enkelmann, V., Wenz, G.: unpublished results
74. Tieke, B., Bloor, D.: Makromol. Chem. *182*, 133 (1981)
75. Tieke, B., Bloor, D., Young, R. J.: J. Mater. Sci. *17*, 1156 (1982)
76. Galiotis, C., Young, R. J., Ando, D. J., Bloor, D.: Makromol. Chem., in press
77. Kiji, J., Kaiser, J., Wegner, G., Schulz, R. C.: Polymer *14*, 433 (1973)
78. Kobelt, D., Paulus, E. F.: Acta Cryst. B *30*, 232 (1974)
79. Enkelmann, V.: Acta Cryst. B *33*, 2842 (1977)
80. Bloor, D., Koski, L., Stevens, G. C., Preston, F. H., Ando, D. J.: J. Mater. Sci. *10*, 1678 (1975)
81. Aimé, J. P.: Thèse. Osay: 1983
82. Lando, J. B.: private communication
83. Enkelmann, V., Lando, J. B.: Acta Cryst. B *34*, 2352 (1978)
84. Apgar, P. A., Yee, K. C.: Acta Cryst. B *34*, 957 (1978)
85. Enkelmann, V.: Habilitationsschrift. Freiburg: 1983
86. Wegner, G.: Makromol. Chem. *154*, 35 (1972)
87. Garito, A. F., McGhie, A. R., Kalyanaraman, P. S.: Kinetics of solid state polymerization of 2,4-hexadiyne-1,6-diol bis (p-toluene sulfonate). In: Molecular metals. Hatfield, W. E. (ed.). New York: Plenum Press 1979, pp. 255–260
88. Kröhnke, C.: Dissertation. Freiburg: 1979
89. Kröhnke, C., Enkelmann, V., Wegner, G.: Chem. Phys. Lett. *71*, 38 (1980)
90. Patel, G. N., Chance, R. R., Turi, E. A., Khanna, Y. P.: J. Amer. Chem. Soc. *100*, 6644 (1978)
91. Grimm, H., Axe, D., Kröhnke, C.: Phys. Rev. B *25*, 1709 (1982)
92. Robin, P.: Thèse. Orsay: 1980
93. Albouy, P. A.: Thèse. Orsay: 1982
94. Albouy, P. A., Patillon, P. A., Pouget, J. P.: Mol. Cryst. Liq. Cryst. *95*, 239 (1983)
95. Robin, P., Pouget, J. P., Comes, R., Moradpour, A.: J. de Phys. *41*, 415 (1980)
96. Bloor, D., Kennedy, R. J., Batchelder, D. N.: J. Polym. Sci., Polym. Phys. Ed. *17*, 1355 (1979)
97. Batchelder, D. N., Bloor, D.: J. Polym. Sci., Polym. Phys. Ed. *17*, 569 (1979)
98. Voigt, W.: Lehrbuch der Kristallphysik. Berlin: Teubner 1910
99. Reuss, A.: Angew. Math. Mech. *9*, 49 (1929)

100. Holliday, L.: Composite Materials. New York: Elsevier 1966
101. Leyrer, R. J., Wettling, W., Wegner, G.: Ber. Bunsenges. Phys. Chem. *84*, 880 (1980)
102. Baughman, R. H.: J. Chem. Phys. *68*, 3110 (1978)
103. Lochner, K., Hinrichsen, T., Wolfberger, W., Bässler, H.: Phys. stat. sol. (a) *50*, 95 (1978)
104. Lochner, K., Bässler, H., Hinrichsen, T.: Ber. Bunsenges. Phys. Chem. *83*, 899 (1979)
105. Bloor, D., Ando, D. J., Hubble, C. L., Williams, R. L.: J. Polym. Sci., Polym. Phys. Ed. *18*, 779 (1980)
106. Bloor, D., Ando, D. J., Fisher, D. A., Hubble, C. L.: Solid state reactivity of some bis (aromatic sulphonate) diacetylenes. In: Molecular metals. Hatfield, W. E. (ed.). New York: Plenum Press 1979, pp. 249–253
107. Bässler, H.: Adv. Polym. Sci., in press
108. Baughman, R. H., Chance, R. R.: J. Chem. Phys. *73*, 4113 (1980)
109. Patel, G. N.: J. Polym. Sci., Polym. Lett. Ed. *16*, 607 (1978)
110. Patel, G. N., Chance, R. R., Witt, J. D.: J. Chem. Phys. *70*, 4387 (1979)
111. Patel, G. N., Walsh, E. K.: J. Polym. Sci., Polym. Lett. Ed. *17*, 203 (1979)
112. Wenz, G., Wegner, G.: Makromol. Chem., Rapid Commun. *3*, 231 (1982); Mol. Cryst. Liq. Cryst. *96*, 99 (1983)
113. Plachetta, C., Schulz, R. C.: Makromol. Chem., Rapid Commun. *3*, 815 (1982)
114. Wenz, G., Müller, M. A., Schmidt, M., Wegner, G.: Polymer, in press
115. Wenz, G.: Dissertation. Freiburg: 1983
116. Sixl, H.: Adv. Polym. Sci., in press
117. Enkelmann, V., Leyrer, R. J., Schleier, G., Wegner, G.: J. Mater. Sci. *15*, 168 (1980)
118. Bloor, D., Preston, F. H.: Phys. stat. sol. (a) *39*, 607 (1977)
119. Robin, P., Pouget, J. P., Comes, R., Moradpour, A.: Chem. Phys. Lett. *71*, 217 (1980)
120. Fukui, M., Sumi, S., Hatta, I., Abe, R.: Jap. J. Appl. Phys. *19*, L 559 (1980)
121. Cailleau, H., Moussa, F., Mons, J.: Solid State Comm. *31*, 521 (1979)
122. Kiess, H., Clarke, R.: Phys. stat. sol. (a) *49*, 133 (1978)
123. Xiao, D. Q., Ando, D. J., Bloor, D.: Mol. Cryst. Liq. Cryst. *95*, 201 (1983)
124. Kennedy, R. J., Chalmers, I. F., Bloor, D.: Makromol. Chem., Rapid Commun. *1*, 357 (1980)
125. Lacey, R. J., Williams, R. L., Kennedy, R. J., Bloor, D., Batchelder, D. N.: Chem. Phys. Lett. *83*, 65 (1981)
126. Bloor, D., Chalmers, I. F., Kennedy, R. J., Motevalli, M.: Mol. Cryst. Liq. Cryst. *95*, 215 (1983)
127. Williams, R. L., Ando, D. J., Bloor, D., Hursthouse, M. B.: Polymer *21*, 1269 (1980)
128. Hädicke, E., Mez, E. C., Krauch, C. H., Wegner, G., Kaiser, J.: Angew. Chem. *83*, 253 (1971)
129. Berkovitch-Yellin, Z., Leiserowitz, L.: J. Amer. Chem. Soc. *97*, 5627 (1975); Acta Cryst. B *33*, 3657 (1977)
130. Irngartinger, H., Jäger, H. U.: Angew. Chem. *88*, 615 (1976)
131. Melveger, A. J., Baughman, R. H.: J. Polym. Sci., Polym. Phys. Ed. *11*, 603 (1973)
132. Baughman, R. H., Witt, J. D., Yee, K. C.: J. Chem. Phys. *60*, 4755 (1974)
133. Lewis, W. F., Batchelder, D. N.: Chem. Phys. Lett. *60*, 232 (1979)
134. Cottle, A. C., Lewis, W. F., Batchelder, D. N.: J. Phys. C *11*, 605 (1978)
135. Mitra, V. K., Risen, Jr., W. M., Baughman, R. H.: J. Chem. Phys. *66*, 2731 (1977)
136. Chance, R. R., Baughman, R. H., Müller, H., Eckhardt, C. J.: J. Chem. Phys. *67*, 3616 (1977)
137. Iqbal, Z., Chance, R. R., Baughman, R. H.: J. Chem. Phys. *66*, 5520 (1977)
138. Exarhos, G. J., Risen, Jr., W. M., Baugham, R. H.: J. Amer. Chem. Soc. *98*, 481 (1976)
139. Boudreaux, D. S., Chance, R. R.: Chem. Phys. Lett. *51*, 273 (1977)
140. Bloor, D., Hubble, C. L.: Chem. Phys. Lett. *56*, 89 (1978)
141. Kuhn, H.: Fortschr. Chem. Org. Naturstoffe *16*, 169 (1958); *17*, 404 (1959)
142. Porod, G.: Monatsh. Chem. *80*, 251 (1949)
143. Kratky, O., Porod, G.: Rec. Trav. Chim. *68*, 1106 (1949)
144. Lim, K. C., Fincher, L. H., Heeger, A. J.: Phys. Rev. Lett. *50*, 1934 (1983)
145. Berlinsky, A. J., Wudl, F., Lim, K. C., Fincher, C. R., Heeger, A. J.: Theory of the rod-to-coil transition in polydiacetylene. Preprint 1983; Mol. Cryst. Liq. Cryst., in press
146. Müller, M. A., Schmidt, M., Wegner, G.: Makromol. Chem., Rapid Commun., in press

147. Tomkiewicz, Y., Engler, E. M., Scott, B. A., La Placa, S. J., Brom, H.: Doping organic solids —
     its uses to probe and to modify electronic properties. In: Molecular metals. Hatfield, W. E. (ed.).
     New York: Plenum Press 1979, pp. 43–49
148. Kitaigorodskii, A. I.: Molecular crystals and molecules. New York: Academic Press 1973
149. Enkelmann, V.: Makromol. Chem. *179*, 2811 (1978)
150. Enkelmann, V.: J. Mater. Sci. *15*, 951 (1980)
151. Ando, D. J., Bloor, D., Tieke, B.: Makromol. Chem., Rapid Commun. *1*, 385 (1980)
152. Kröhnke, C.: Diplomarbeit. Freiburg: 1977
153. Patel, G. N., Miller, G. G.: Copolymerization of diacetylenes in the crystalline solid state —
     a method for recording latent finger prints. Preprint: 1976
154. Yee, K. C.: J. Org. Chem. *44*, 2571 (1979)
155. Sixl, H.: private communication
156. Chance, R. R., Yee, K. C., Baughman, R. H., Eckhardt, H., Eckhardt, C. J.: J. Polym. Sci.,
     Polym. Phys. Ed. *18*, 1651 (1980)

H.-J. Cantow (Editor)
Received January 18, 1984

# Author Index Volumes 1–63

# Subject Index